John Cresson Trautwine

A Method of Calculating the Cubic Contents of Excavations and Embankments

By the aid of diagrams. Together with directions for estimating the cost of earthwork. Ninth Edition

John Cresson Trautwine

A Method of Calculating the Cubic Contents of Excavations and Embankments
By the aid of diagrams. Together with directions for estimating the cost of earthwork. Ninth Edition

ISBN/EAN: 9783337106027

Printed in Europe, USA, Canada, Australia, Japan

Cover: Foto ©berggeist007 / pixelio.de

More available books at **www.hansebooks.com**

A METHOD

OF

CALCULATING THE CUBIC CONTENTS

OF

EXCAVATIONS AND EMBANKMENTS,

BY THE AID OF DIAGRAMS.

TOGETHER WITH

DIRECTIONS FOR ESTIMATING THE COST OF EARTHWORK.

BY

JOHN C. TRAUTWINE,
CIVIL ENGINEER.

NINTH EDITION.

REVISED AND ENLARGED
By JOHN C. TRAUTWINE, Jr., C. E.

NEW YORK:
JOHN WILEY & SONS,
15 Astor Place.
1887.

PREFACE.

THE graphic method here given for finding the depths of level cross sections equivalent in area to given sections with sloping ground surface, originated with the author many years since, and was first published by him in 1851.

The diagrams are accompanied by tables of contents of level cuttings; and directions are given for calculating irregular sections. The tables were prepared with the greatest care; and have undergone so thorough a revision as to leave scarcely a doubt of their entire reliability. The remarks on the measurement of earth-work are supplemented by a method for estimating its cost, in which the author follows that proposed by the late Ellwood Morris, C. E., of Philadelphia, in the Journal of the Franklin Institute in 1841.

Remarks on the operation of steam excavators, and rules for estimating the cost of moving earth by wheeled and drag scrapers and by locomotive and cars, have been added. They are taken from the later editions of the author's "Civil Engineer's Pocket-Book."

A new method of constructing tables of level cuttings, suggested by Mr. John R. Hudson, C. E., has been incorporated in the present work by his permission.

J. C. T., Jr.

June, 1887.

A METHOD

OF

CALCULATING THE CUBIC CONTENTS

OF

EXCAVATIONS AND EMBANKMENTS.

THERE is but one correct principle upon which to calculate the cubic contents of excavations and embankments; and that is, the *Prismoidal Formula*, or Rule; which is as follows:

Add together the areas of the two parallel ends of the prismoid, and four times the area of a section half-way between and parallel to them; and multiply the sum by one-sixth of the length of the prismoid, measured perpendicularly to its two parallel ends.

Since, in railroad measurements, the prismoids are generally 100 feet long, it becomes easier in practice to multiply the sum of the areas in square feet, by 100, (by merely adding two ciphers,) and to divide the product by 6; which amounts to the same thing as multiplying their sum by $\frac{1}{6}$th of 100 feet.

The very extended application of the prismoidal formula to other solids than such as are commonly understood by the term "prismoids," was first shown by Mr. Ellwood Morris, Civil Engineer, in a paper published in the Journal of the Franklin Institute, in 1840.

It embraces all parallelopipeds, pyramids, prisms, cylinders, cones, wedges, &c., whether regular or irregular, right or oblique; together with their frustums, when cut by planes parallel to their bases; in a word, *any solid whatever, which has two parallel ends, connected together by longitudinally unwarped surfaces,* whether plane or curved. It also applies to spheres, hemispheres, spheroids, paraboloids, &c.

In the cylinder and cone, the sides may be considered as consisting of an infinite number of infinitely narrow planes, unwarped longitudinally. In railroad cuttings, it rarely happens that the surface planes lying between two consecutive cross sections, 100 feet apart, are absolutely unwarped; yet, for practical purposes, they may very frequently be assumed to be so. When much warped, the cross sections must be taken closer together than 100 feet. *Upon a strict attention to this precaution depends the accuracy of earthwork measurements; the entire principle of which is embraced in the foregoing remarks.* No practicable method is perfectly accurate. All we can do in actual practice is, to take our stations at distances so near together that the intermediate solid shall be *very nearly* a prismoid, and then calculate it *as if a true* prismoid.

There are generally two circumstances under which it is necessary to compute the cubic contents on a public work; viz.: first, after a preliminary survey of one or more *trial lines;* for the purpose of determining *approximately* their actual, or comparative costs; and, second, after the final adoption and staking out of the determined route, in order to know more precisely the amount of work to be done.

The measurements for the latter are performed with more care, and attention to detail, than those of the former, inasmuch as upon them depend the payments to be made to the person who executes the work. They, moreover, involve considerations which cannot be attended to during a preliminary survey, without incurring an expenditure of time and labor, more than commensurate with the importance of the result.

When the ground is *level* transversely of the line of survey, there is no difficulty whatever in ascertaining the contents from a table of *level-cuttings,* previously calculated; but when it is *inclined* or *irregular* transversely, more or less calculation is required.

The following method by diagrams will we trust, be found to facilitate the operations in the last-named cases. It dispenses with a great deal of calculation; and is, therefore, comparatively free from liability to errors arising from that source.

METHOD OF USING THE DIAGRAMS.

The construction of the diagrams is extremely simple, notwithstanding that, at first sight, they appear somewhat complex. They are but few in number, since any particular road will generally require but three or four, which may be prepared by one person in a few days. Before proceeding to explain the manner of drawing them, we will give one or two examples of their use, that the reader may see the object aimed at, and to what extent it is attained.

Example 1. Suppose that with a roadway 28 feet wide, and with sideslopes of 1½ to 1, the cutting at a certain station is 20 feet deep on centre line; and that the ground, instead of being level transversely, inclines at an angle of 15° with the horizon.

Turn to the diagram, Plate IX., for a roadway 28 feet wide, with sideslopes of 1½ to 1: place a finger on the centre line, at the height of 20 feet, and run it along up the curved line which commences at that point, until it strikes the inclined line marked 15°. It will be seen at once that the two coincide at the height of 22·8 feet: *and this is the depth of the equivalent level cutting, which would have precisely the same area as the section under consideration.*

All such cases may therefore be instantly, and without any calculation whatever, reduced to others of *equivalent level cuttings.*

This constitutes the main feature of the principle involved in the diagrams.

Had the depth been 20·3, or other decimal of a foot, the proceeding would have been the same as with the 20 feet; and the equivalent level cutting would be found on the inclined line 15°, at the distance of ·3 of a foot (estimated by eye) above the curved line 20.

Example 2. Using the same diagram; let the depth of cutting be 2 feet, and the transverse slope of the ground 20°. Here, placing a finger on the centre line, at the height of 2 feet, and running it along the curved line commencing at that point, it will be found that before reaching the inclined line of 20°, it encounters the *dotted* curved line drawn near the bottom of the diagram. When this occurs, we know that the ground-slope cuts the roadway, forming a cross section partly in excavation and partly in embankment, as in fig. 9.

This is a most useful check; for in such cases, the contents cannot be obtained by means of the diagram*; but recourse must be had to a figure of the section drawn for the purpose; as must also be the case when the ground is *irregular* transversely. A simple method of proceeding, in all such cases, will be given further on.

On the page opposite each diagram, is a table of cubic yards for level cuttings, and for lengths of 100 feet. By means of these tables, the cubic contents may at once be taken out, when the equivalent level cuttings at both ends of a station are equal, and the ground-slope between them uniform: but if the equivalent level cuttings at the two ends of the station are unequal, then the prismoidal rule must be applied; thus,

Suppose the equivalent level cutting at one end to be 20 feet, and at the other 25 feet, and the intervening ground-slope uniform. Then the equivalent level cutting at a point half-way between them would

---

* The equivalent centre height given by the diagram in such cases is that of a level cutting whose area of cross section is equal to the *difference* between that of the embankment $a e o a$, fig. 9, and that of the excavation $a d b a$.

be 22½ feet. Therefore, the cubic content will be equal to one-sixth of the sum of those corresponding to each of the two end depths and of four times that of the centre depth; that is,

Cubic content by table 9, for 20 feet depth, = 4296 cubic yards,
" " " " 25 " = 6065 " "
Four times " " 22½ ⎫ " = 20584 " "
or 4 times 5146 ⎭

$$\overline{6)30945}$$
Cubic yards contained in the station, = 5157·5

These tables are carried to depths or heights of 60 feet; but in the subsequent table No. 15, they are extended to 170 feet. As these extended quantities will be but seldom referred to, they are calculated only to whole feet; but the amount corresponding to any fraction of a foot may be found with sufficient accuracy for practice, by simple proportion.

It will be perceived that, instead of the *areas* corresponding to the different depths of cutting, or heights of filling, our tables give the *cubic yards* corresponding to those areas, for lengths of 100 feet. For the purposes of calculating cubic contents, these solidities may evidently be used instead of the areas. Table 17 gives the cubic yards in a prism 100 feet long and of any given area of cross section from 1 square foot to 1000 square feet. Its use will be shown further on.

For rough preliminary estimates of trial lines, the labor may be much reduced by taking from the tables, the cubic content corresponding to the *average* of the equivalent level cuttings at the two ends. This mode is not mathematically correct, and should never be resorted to for final estimates; but it will be sufficiently approximate (always a little deficient) for such cases as occur in ordinary cuttings and fillings; even if the depths or ground slopes at one end differ from those at the other, provided the depths do not differ more than about 5 feet, nor the ground slopes more than about 5°, and provided said slopes are in the same direction.

For instance, in the foregoing example, the correct contents of the station 20 feet deep at one end, and 25 feet at the other, with the same ground slope at each end, were found to be 5157·5 cubic yards; while, by this approximating mode, the contents of an *average* level depth of 22½ feet, would be 5146 cubic yards; or but 11½ yards less than the truth.

Or, for true prismoids, (or even otherwise within the above limits of no greater differences than 5 feet in depth and 5° in slope at the two ends of a 100 feet station, the slopes being in the same direction), we may add together the tabular contents corresponding to the two equivalent level depths at the ends of the station, and divide their sum by 2. The content thus found will not be as approximate, however, as that by the first method; but will be *too great* by precisely *twice* the quantity that the other is *too small*. Thus, in the foregoing example, we should have for a true prismoid,

Depth.	Cubic yards.
20	4296
25	6065
	2)10361
	5180 cubic yards = approx. content,

or 23 yards in excess of the true content, $5157\frac{1}{2}$ yards; or twice the deficiency ($11\frac{1}{2}$ yards) of the preceding method.

These examples merely show that in railroad work, and within limits of frequent occurrence, we may calculate the content of a true prismoid by either of these approximate modes, with sufficient accuracy for rough preliminary, or comparative estimates. We have in neither instance given the actual content of a solid whose transverse slopes differ at its two ends. Said content would be farther from the truth than in our examples; where, by the first method, the error is but 1 yard in about 450; and in the second, 1 in about 225; whereas the average of a number of stations in which the slopes at the two ends differ on an average $2\frac{1}{2}°$, and in no case more than $5°$, would probably be in error by about 1 yard in 100 too little, by the first method; and 1 yard in 50 too much, by the second.

For *final* estimates, however, we should make our stations so short that the ground surface of the included solid may be considered unwarped longitudinally, and then use the prismoidal rule.

PRINCIPLE ON WHICH THE METHOD IS BASED,

To find the sides of a triangle of which only the area and the angles are given.

RULE.—In any plane triangle, as the product of the sines of any two of the angles, is to the sine of the remaining angle, so is twice the area of the triangle, to the square of the side lying between the two angles first taken.

Fig. 1.

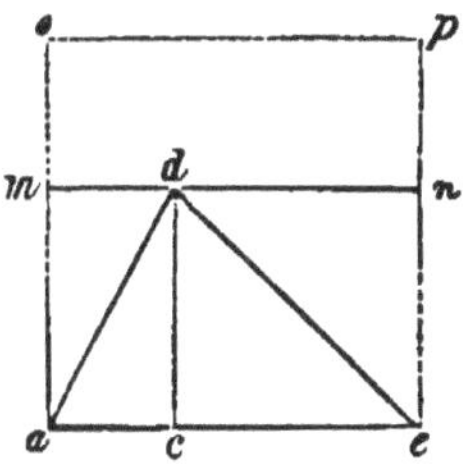

Demonstration.—Let $a\,d\,e$ be a triangle, in which we have given, its area, and its three angles: it is required to find any side, as $a\,e$.

By trigonometry we have the two following proportions:—

Sine of a : Radius (1) :: $d\,c$: $a\,d$; also

Sine of e : Sine of d :: $a\,d$: $a\,e$.
the angle opp. $a\,d$ ⋮ the angle opp. $a\,e$

By multiplication of these two proportions, we have—

Sine of a × Sine of e : Sine of d :: $d\,c × a\,d$: $a\,e × a\,d$; or, leaving out the factor ad, common to the last two terms, Sine of a × Sine of e : Sine of d :: dc : ae.

But, as dc : ae :: $dc \times ae$: ae^2.

Again, $dc \times ae =$ twice the area of the triangle ade.

Hence we have: Sine of a × Sine of e : Sine of d :: 2 area : ae^2. Q. E. D., and the square root of $ae^2 = ae$, the required side.

Now, let $nmcb$, fig. 2, be the level cutting equivalent, or equal, to the side-hill cutting $nmed$. From the point of intersection at i, draw ia.

Fig. 2.

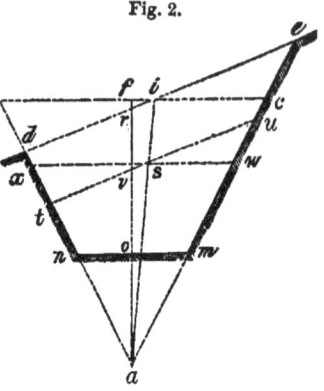

Then, if tu represent any other side-hill cutting parallel to de, we have only to draw the horizontal line xw, through s, in order to obtain the equivalent level cutting $nmwx$.*

The point required is to find fi, the distance to be laid off on the horizontal line fc, in order to draw ia.

To do this, draw a sketch as $bcmn$, fig. 2, either by scale or not, showing a level cutting (or filling, as the case may be) of any center depth fo at pleasure, say 10, 20, or 30, &c., feet. To this sketch add the triangle mna. Calculate the area of $bcmn$; also the height (ao) and the area of the triangle mna. Then on the sketch draw a line de, representing that ground-slope (5°, or 10°, &c.) for which the distance fi is being sought; and crossing the center-line fo, as at r.

Then *assume* that the area $demn$ is equal to the known area of the level cutting $bcmn$; and that consequently the area of the triangle dea is equal to the known area of the triangle bca.

Next, from this area of the triangle dea, find its side ae, as per rule on page 9, thus:

$$\text{Sine of } a \times \text{Sine of } e : \text{Sine of } d :: \text{Twice the area of } dea : ae^2.$$

Then find ra, thus:

$$\text{Sine of } era \text{ opp. given side } ae : \text{Sine of } e \text{ opp. req'd. side } ra :: ae : ra.$$

Then $fa - ra = rf$.

Also, the angle eic, representing the ground slope, is equal to the angle fir; and considering fi as a radius, and rf as a tangent to the angle fir, we have—

Tang. of fir, (or ground slope,) : Rad, or 1 :: rf : fi.

Then as af : fi :: 1 : fi when af is assumed as unity, in preparing a working diagram, and as in the next table.

To save the trouble of calculating these distances fi, we have extended the table, p. 11, to all side-slopes likely to occur in practice.

---

* The demonstration of this, on the principle of similar triangles, is so simple as not to require insertion.

METHOD OF PREPARING THE DIAGRAMS.

We will now proceed to describe the mode of preparing the diagrams, for any width of roadway, and for any side-slope whatever.

Draw a vertical line $a\,b$, fig. 3, of any given length at pleasure. (One foot decimally divided; or $12\frac{1}{2}$ inches, divided into $\frac{1}{8}$ths of an inch, or 10 inches divided into $\frac{1}{10}$ths of an inch, will generally be found convenient.) Call the length of this line unity, or 1. It represents the usual centre-line of levels, or of cuttings and fillings.

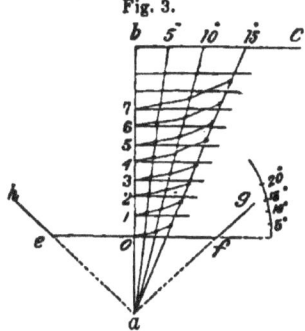

Fig. 3.

From the upper end of this line draw $b\,c$, at right angles to it; and from b towards c, lay off and number the distances $b5°$, $b10°$, $b15°$, &c., contained in the following table; using as a scale the length $a\,b$, as 1 or unity, divided into tenths and hundredths.

For example; if the side-slopes $h\,e$, $g\,f$, of the excavation or embankment, are $\frac{1}{4}$ to 1, lay off (without any regard to the width of roadway,) the distances in the upper line of the table; if 1 to 1, those in the 3d line, &c. This done, the scale of $a\,b$, as unity, will be of no further use.

Distances on $b\,c$, intermediate of those in the table, may be inserted with sufficient accuracy by eye.

Table of Distances f i, Fig. 2; or b 5°, b 10°, &c., Fig. 3, to be laid off on the Horizontal Line b c, Fig. 3; the Center-Line a b, Fig. 3, being assumed as Unity, or 1.

Side Slopes.											
¼ to 1, or 75° 58'	10° ·005	20° ·011	25° ·014	30° ·017	35° ·022	40° ·026	45° ·032	50° ·038	55° ·046	60° ·057	65° ·073	70° ·099
½ to 1, or 63° 26'	5° ·011	10° ·022	15° ·034	20° ·046	25° ·058	30° ·072	35° ·090	40° ·110	45° ·132	50° ·165	53° ·189	55° ·211
1 to 1, or 45°	5° ·044	10° ·089	15° ·136	18° ·167	20° ·188	23° ·222	25° ·247	28° ·288	30° ·318	33° ·369	36° ·431	39° ·510
1¼ to 1, or 39° 40'	5° ·068	10° ·138	13° ·184	15° ·214	18° ·264	20° ·300	23° ·358	25° ·401	28° ·476	30° ·530	32° ·600	34° ·685
1½ to 1, or 33° 42'	5° ·097	8° ·158	10° ·201	13° ·267	15° ·314	18° ·390	20° ·445	22° ·506	24° ·574	26° ·652	27° ·696	28° ·747
2 to 1, or 26° 34'	3° ·106	5° ·175	8° ·295	10° ·363	12° ·447	14° ·533	16° ·629	18° ·739	19° ·798	20° ·865	21° ·936	22° 1·017
2½ to 1, or 21° 48'	2° ·112	4° ·226	6° ·340	8° ·454	10° ·582	12° ·719	14° ·875	16° 1·056				
3 to 1, or 18° 26'	2° ·160	4° ·322	6° ·486	8° ·660	10° ·858	12° 1·080	14° 1·349					

From the points 5°, 10°, 15°, &c., on the line $b\ c$, (and from the subdivisions of single degrees between them, *as shown in the working diagrams*,) draw lines to a. From a upwards, set off, by any scale at pleasure, (about ¼th inch to a foot will be found convenient,) the distance $a\ o$, which is the height of the triangle $e f a$, formed by the prolongation of the side-slopes $g f$, and $h\ e$ to a; $e f$ representing the width of the roadway, whatever it may be, on the same scale.

It is not necessary actually to draw $h\ a$, $g\ a$ and $e f$, as we may set off $a\ o$, by recollecting that if the side-slopes are

¼ to 1, then $a\ o$ will be 4 times $o f$, (the half width of roadway.)
½ to 1, " " twice $o f$.
1 to 1, " " equal to $o f$
1¼ to 1, " " ·8 of $o f$.
1½ to 1, " " ⅔ of $o f$.
2 to 1, " " ½ of $o f$.
2½ to 1, " " ·4 of $o f$.
3 to 1, " " ⅓ of $o f$.

Beginning at o, divide the vertical or centre line $o\ b$, by the same scale into feet; numbering them 1, 2, 3, &c., from o upwards; and from the points of division 1, 2, 3, &c., draw horizontal lines parallel to $b\ c$, as shown in fig. 3

From o as a centre, lay off with a protractor, the several angles of transverse ground-slope as shown by the arc in fig. 3. Angles higher than 20° will seldom be required.

In fig. 3, the inclined lines, and also the angles on the arc, are, for convenience, numbered only for every 5°; but in a *working* diagram they should be taken nearer together, for instance, every 2° to 3°.

Lay a parallel ruler from o to 5° on the arc, and mark with a dot the point of intersection on the inclined line a 5°; then keeping the ruler in the same position, move it upwards along $o\ b$, stopping at every division of 1 foot, and making corresponding dots on the inclined line a 5°, as in fig. 3, continuing to such a height on the centre line as will include the greatest cutting or filling to be calculated by the diagram.

Then lay the parallel ruler from o to 10° on the arc, and mark with a dot, the point of intersection on the inclined line a 10°; then keeping the ruler in the same position, move it upwards along $o\ b$, stopping at every division of 1 foot, making corresponding dots on the inclined line a 10°.

Then lay the parallel ruler from o to 15° on the arc, and proceeding as before, make corresponding dots on the inclined line a 15°, and so on up to as high an angle as will equal the greatest transverse slope of the ground which occurs on the work to be calculated by means of the diagrams.

Finally, connect the corresponding dots on the several inclined lines, forming thereby a series of curved guide-lines, as in fig. 3 and in our working diagrams. The diagram is now ready for use, for all cases of ground-slope which do not intersect the roadway, thereby forming a section partly in excavation, and partly in embankment, as shown in fig. 9.

In order that the diagram itself may inform us when this is the case, the dotted curve shown near the bottom of the working diagrams is added. It is prepared as follows:

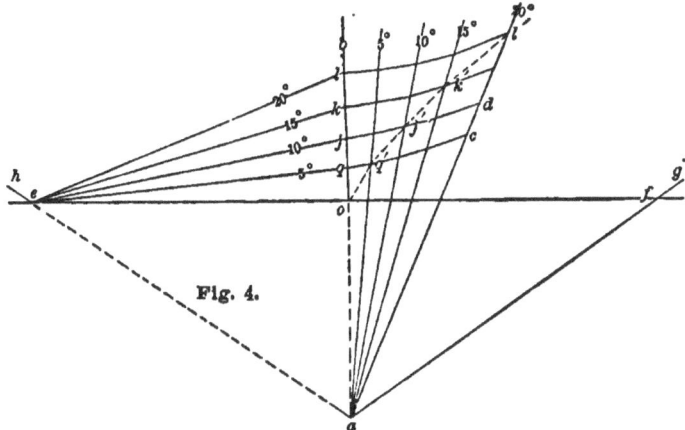

Fig. 4.

From e or from f, fig. 4, where the side slope $h\,e$ or $g\,f$ intersects the roadway $e\,f$, lay off, toward the centre line $a\,b$ and above $e\,f$, the same angles (5°, 10°, 15°, etc.) of transverse ground slope as were laid off from o in fig. 3; draw the lines $e\,q$, $e\,j$, $e\,k$, etc., and mark the points q, j, k, etc., where they intersect the centre line $a\,b$. Now, if a cutting of 5° transverse ground slope has a less centre depth than $o\,q$, its ground slope (parallel to and below $e\,q$) must evidently intersect the roadway $e\,f$; as will also one of 10° ground slope with a less centre depth than $e\,j$; and so on. Hence the radial line a 5° must not be used for a cutting of less centre depth than $o\,q$; nor a 10° for one of less centre depth than $o\,j$. We therefore mark radial line a 5° at the point q', where it would be cut by a curved guide-line $q\,c$ starting from q; radial line a 10° at j', where it would be cut by a curve $j\,d$ starting from j; and so on. Join the marks q', j', k', etc., so made, thus forming the cautionary curve o, q', j', k', l', etc.; and the diagram is finished.

The working drawings which we have given, are on a very small scale, for convenience of insertion in this volume; yet, although the curved lines are drawn straight across several divisions of the inclined lines, (generally five of them,) they will rarely be found, in operating with them, to differ as much as $\frac{1}{4}$th of a foot from the truth in the depth of the equivalent level cutting.

They are adapted to single and double track embankments, 14 and 24 feet wide on top, and with side-slopes of $1\frac{1}{2}$ to 1; and to single and double track excavations, 18 and 28 feet wide at bottom, with side-slopes from 1 to 1, up to 2 to 1; gauge 4 feet 8½ inches.

The widths for 4 feet 8½ inches gauge will rarely differ more than about 2 feet from those for which the diagrams have been prepared. The most mistaken economist would hardly venture to make them more than 2 feet less; nor do we conceive that any great advantage would attend making them more than about 2 feet greater, for a gauge of 4 feet 8½ inches, with cars of the usual 9½ feet extreme width, from out to out of cornice.

With cars of 11 feet extreme width, embankments should not be less than 15 and 27 feet wide; nor cuts less than 19 and 31 feet. We consider *all* the foregoing widths of *embankment* sufficient, but would recommend an addition of 2 or 3 feet to all the *cuts*, except when in rock, to allow for wider and deeper side-ditches than are usually made.

No diagrams accompany the tables of level cuttings for side-slopes steeper than 1 to 1. With very steep side-slopes, such as ¼ to 1, which is used only for rock, the traverse inclination of the ground rarely affects the quantity of material to an important extent. Still, on every work on which much rock-cutting occurs, a diagram should be prepared for the purpose.

The diagrams and tables given in this volume may be used for any greater or less widths of roadbed than those to which they are especially adapted. In other words, it is not at all *necessary* to prepare new ones for every width of roadway.

Suppose, for instance, we wish to use diagram, Plate 1, for an embankment, $m\,n\,h\,q$, fig. 4½, having side-slopes of 1½ to 1, as in the diagram; but with a roadbed $m\,n$ 16 feet in width, instead of $e\,f$ (14 feet), for which latter width the diagram was prepared.

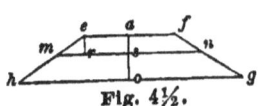
Fig. 4½.

Find the vertical distance $a\,s$, between the 14 feet roadbed $e\,f$, and the 16 feet one $m\,n$, and add it *mentally* to each height $o\,s$, of the level section $m\,n\,h\,q$, as found by the diagram when taking out from the table the number of cubic yards. By this means obtain the contents of the embankment $e\,f\,h\,q$. Next, from these contents so obtained for the entire length of the embankment, subtract that corresponding to the depth $a\,s$, taken from the same table and multiplied by the number of stations of 100 feet comprised in the length of the embankment, or excavation, as the case may be. As $m\,n$ in this instance is 16 feet, and $e\,f$ 14 feet, it follows that $r\,m$ is $\frac{16-14}{2}=1$ ft.; and (the slope being 1½ horizontal to 1 vertical) $a\,s\,(=e\,r)$ is $\frac{1}{1\frac{1}{2}}=\cdot 6667$ ft. In such cases use mentally the nearest decimal of a foot in working with the tables, inasmuch as they are calculated only for single decimals of a foot. Thus, in this case add ·7 of a foot to every height $s\,t$ of the embankment $m\,n\,o\,p$.

A separate diagram is absolutely required therefore only for each side-slope: and such a diagram may be used indifferently for either excavation or embankment, provided the two have the same side-slope, and for any width of base or roadway whatever.

Remark. In using a working diagram, however, for a width of roadway different from that for which it was originally made, a new dotted curved line corresponding to the new width must be first laid down upon it, prepared by the directions given at fig. 4. This, however, can be done in a few minutes.

In preparing working diagrams, they should be made with reference to the greatest depths of cut or fill that occur on the route. Ours extend only to 60 feet, for convenience of insertion in this volume. A scale of about ⅛ to ¼ of an inch to a foot will be quite sufficient.

IRREGULAR SECTIONS.

Fig. 5 illustrates the construction of Trautwine's Cross-section Sheet,* upon which irregular cross sections may be drawn, and thus, by inspection, be reduced to equivalent level or sloping sections, and the latter may then be reduced to equivalent level sections by means of diagrams similar to Plates I to X of this work.

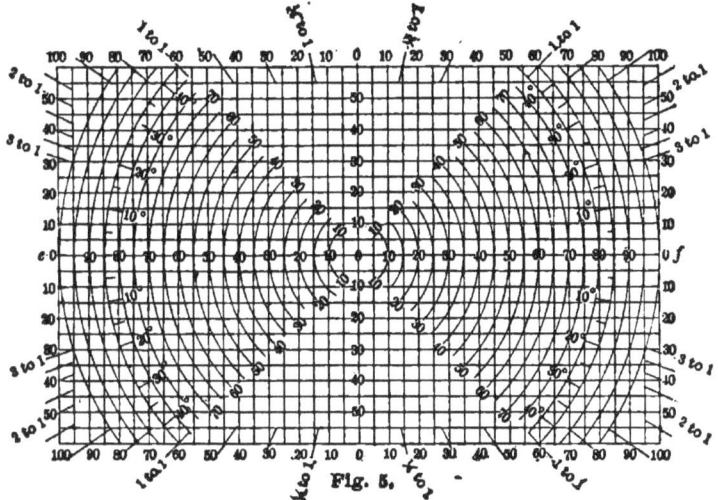

Fig. 5.

The squares in fig. 5 represent squares of half an inch in the working sheet. On said sheet these are divided into smaller squares $\frac{1}{10}$ of an inch on a side, the scale being $\frac{1}{10}$ of an inch to a foot.

The given cross-section is to be drawn lightly upon the printed sheet, with the centre of the roadway at the centre *o* of the sheet,

---

* Published by John Wiley & Sons, New York, price, 25 cts. per sheet; $5.00 per quire.

and with the roadway coinciding with the horizontal line ef. The width of roadway is laid off on ef, the squares ruled on the sheet serving as a scale of $\frac{1}{10}$ of an inch to a foot. The side slopes are drawn by means of the slope lines printed around the edge of the diagram. Thus, if the side slopes are 2 to 1, lay a parallel ruler from o to the short line marked 2 to 1 near the right hand upper corner of the diagram. (For greater correctness the ruler should be long enough to reach at the same time to the corresponding line marked 2 to 1 at the lower left hand corner of the sheet.) Then move the ruler, parallel with itself, to the right until it cuts the right hand end of the line just laid off to represent the roadway, and draw the right hand slope.

A similar operation, with the ruler reaching from the 2 to 1 line at the left hand upper corner to that at the right hand lower one, and moved from the centre toward the left, enables us to draw the left hand slope.

The transverse ground slopes are to be laid off upon the cross section sheet in the same way; using the protractor printed upon the sheet, and always laying the parallel ruler first from the centre o to the required degree on the protractor.

One sheet may be made to serve for the calculations of many stations, oy merely drawing in the transverse ground-slopes, *very lightly*, with lead pencil marks, which may be rubbed out as each station is finished.

It is advisable, however, in *very irregular* sections, to represent but two consecutive ones on one sheet; and after having drawn them in ink, and added the numbers of the sections or stations to which they belong, as well as the cubic content comprised between them, to lay them aside for future reference, in case of dispute with the contractor, after the work is commenced.

The method we advise for reducing irregular cross-sections to equivalent regular ones, which may be calculated by means of the working diagrams and tables of level cuttings, is as follows:

CASE 1.—*When the ground slopes differently from the centre each way, as* e d, e a, *in figs.* 6 *and* 7.

Fig. 6.   Fig. 7.

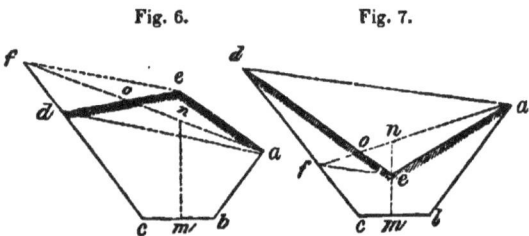

On the prepared paper, fig. 5, from the centre-height or depth e, figs. 6 and 7, and by means of the parallel ruler, and the degrees marked on the sheet, fig. 5, draw the two slopes $e\ d$, $e\ a$, figs. 6 and 7; the inclinations of which are taken from the field slope-book. Draw ef, parallel to ad, and join af. Then is $abcf$ equal to $abcde$, figs. 6 and 7.

PROOF.—The two triangles adf, ade, being on the same base ad, and between the same parallels ad and ef, are equal to each other. Leaving out from each, the triangle ado, which is common to both, we have the triangle dfo, equal to the triangle aeo; and consequently $abcf$ is equal to $abcde$.

Find by means of the parallel ruler and degrees marked on the paper, the slope of af; and with that slope, and the new centre-depth mn, (which is had from the figure by inspection,) use the proper diagram for finding the equivalent level-cutting; and take out the cubic yards from the table.

CASE 2.—*When the ground is very irregular transversely, as in fig. 8.*

Having drawn the figure on the prepared paper, find by trial with a piece of thread, the line ad, which equalizes, as nearly as can be judged by eye, the irregularities above and below it. By means of the parallel ruler, and the degrees on the paper, find the slope of ad; and with that slope, and the new centre-depth mn, (which is had from the figure by inspection,) use the proper diagram for finding the equivalent level-cutting; and take out the cubic yards from the table.

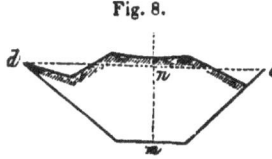

Fig. 8.

CASE 3.—*When the ground-slope intersects the roadway, as in fig. 9.*

Such cases are always detected by the dotted curve line in the working diagrams.

Having drawn the figure on the prepared paper, measure the two bases ab and ao; and also the two perpendiculars to them, cd and ef.

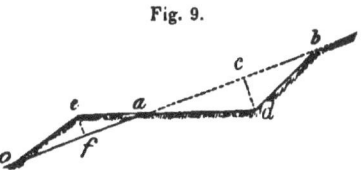

Fig. 9.

Multiply ab by cd, and *half* the product is the area of the triangle of excavation abd. If the triangle of excavation at the other end of the 100 feet station has the same area, the number of cubic yards corresponding to this area for a length of 100 feet will be taken from table 17.

Also multiply ao by ef, and *half* the product will be the area of the triangle of embankment aoe. If the triangle of embankment at the other end of the 100 feet station has the same area, the corresponding number of cubic yards will be taken from table 17.

But if the triangles of excavation, (or those of embankment,) at the two ends, are not of the same area, the prismoidal rule must be employed, as in the case of trapezoidal cross sections of unequal areas.

That is, we must add together the contents corresponding to the two end triangles, and 4 times that corresponding to the triangle half way between them, and divide the sum by 6, for the true content.

The base of the centre triangle of excavation, will be the average between the two bases $a\,b$, fig. 9, at the ends; and its perpendicular, the average between the two perpendiculars $c\,d$, at the ends.

In like manner, the base and perpendicular of the centre triangle of embankment, will be averages of the two end bases $a\,o$, and of the two end perpendiculars $e\,f$.

If from irregularities in the ground, in the direction of the line of the road, it should become necessary to take cross-sections nearer together than 100 feet, only the same proportional parts of the cubic yards must be taken from the tables; and on this account, it is better always, when possible, to subdivide the 100 feet station-distances into such parts as will furnish numbers easy to divide by; thus, if the station be divided into 10, 20, 25, or 50 feet distances, they will furnish respectively the numbers 10, 5, 4, or 2, by which to divide the cubic yards in the tables, all of which are calculated for 100 feet distances.

TO FIND THE DISTANCES OF THE SIDE-STAKES FROM THE CENTRE-STAKE.

For this purpose the circular arcs shown in fig. 5 are printed upon the cross-section sheet. Lay a parallel ruler from the point representing the centre stake to that representing the side-stake, and move it, parallel to itself, until it touches the point O. Note the point where the ruler now cuts the vertical line on which the side stake is drawn. The distance of this point from O is given by the circular arcs, and is equal to the distance between the centre and side-stakes as measured along the sloping ground surface. If a scale is at hand this distance can of course be measured directly with less trouble. The *horizontal* distance between the two stakes is given at once by the ruled lines on the sheet.

This method, however, is applicable only when the ground-slopes are regular, and have been taken with great care. The following method by the level is altogether preferable for general purposes.

It is generally best to note the horizontal side distances; because if a side-stake is accidentally lost after the excavation has been commenced, it is then only necessary to find the centre line of the work, in order to replace it; whereas when the inclined side-distances are used, it becomes necessary to find not only the position of the vertical centre *line*, but also the original *height* of the centre *stake*, to measure from.

USUAL METHOD OF FINDING SIDE-DEPTHS, AND PLACING SIDE-STAKES.

1st. For an Excavation.

The level is placed conveniently for sighting from the same position upon f, c, and d, figs. 11 and 12.

Fig. 11.   Fig. 12.

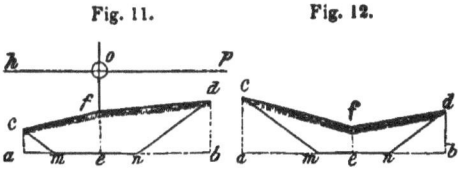

A sight $f o$, is then taken on the target-rod held at f; this sight $f o$, being added to the centre-depth $f e$, gives the height $e o$, of the instrument above $a b$; or the height of the horizontal plane, (represented by $h p_{,}$) through which the line of sight passes as the telescope of the level is swept round on the axis of the instrument.

The height of d above f is then estimated by eye, say at 2 feet; this 2 feet, added to $f e$, gives the *approximate* height of $d b$. Assuming the approximate height $d b$ as the correct one, we find what would be the *horizontal* distance from the centre to d, either by calculation, or from a previously prepared table of horizontal distances. Measure off that distance horizontally towards d, and placing the target 2 feet lower on the target rod, hold it at the end of the measured distance. A sight is then taken with the level, and if it strikes the centre of the target, it proves that the assumed height of d, and the corresponding horizontal distance from the centre of the roadway, were correct; and that the proper spot is found for placing the side-stake d.

It seldom happens that such a coincidence is found at the first trial; at least two trials are generally required; and frequently three, or even four when the ground is extremely irregular.

A very close approximation, however, can always be made by an experienced leveller after the first trial. An error of an inch or two in the position of a side-stake is a matter of no practical importance whatever.

The same operation is performed at c, except that as c is *lower* than f, the target is *raised* on the rod, as far above the sight taken at f as c is estimated to be below f.

2d. For an Embankment.

When putting in side-stakes for an *embankment*, fig. 13, the sight taken on the rod at the centre-stake, is *subtracted* from the centre-height of the embankment, in order to obtain the depth of the instrument below the roadway; and the outer sights, $c a$, $d a$, are to be added to this depth, $s t$, for the side-depths; except when, as in fig. 14, the sight $r y$, on the rod at the centre-stake, *is greater than the height $r x$ of the embankment*, in which case the difference $x y$, between the two, will be the height of the instrument above the roadway; and this difference, $x y$, must then be subtracted from the sights *a c, a d*, for the side-depths, *o c, o d;* all of which is apparent from the figures.

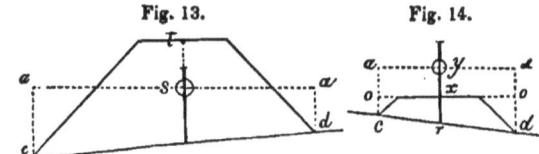

Fig. 13.   Fig. 14.

It is plain that if the height on the target rod at *d*, fig. 11, be subtracted from *e o*, the remainder will be *d b;* and that taken at *c*, subtracted from *e o*, will give *c a*.

These operations give us therefore, at the same time, the heights *d b*, and *c a*; and also the horizontal distances from the centre to each side-stake. All these are at once entered into the proper field-book, to be used in estimating the areas and cubic contents in the office.

The sum of the two horizontal distances manifestly gives the extreme width of the excavation or embankment.

Transverse ground-slopes are obtained in the field, by means of a small slope-level, or *clinometer*, placed upon a rod, 10 or 12 feet long; which, at every station, is laid upon the ground as nearly at right angles to the line of survey, as can be judged by eye. These slope-levels are sold by most instrument-makers. If they were graduated to give the slope in feet per 100 feet, they would be much more convenient.

When the ground is regular transversely, but one slope need be taken; at other times, one or more may be required from the centre-stake each way. The slopes for estimating the final adopted line, need not extend beyond the widths actually occupied by the cuttings and fillings; while those taken in preliminary surveys should comprise a considerable width, as they are, moreover, used in the office for changing the position of the surveyed line, in order to avoid excavation and embankment.

In the following tables of level-cuttings, the left-hand vertical column contains the height or depth of the embankment or excavation, in feet; and the upper horizontal column, the intermediate tenths of a foot. Thus, in table 1, the cubic yards in a station 100 feet long and 10 feet deep, are 1074; for 10·1 deep, 1090; for 10·2 deep, 1107, &c.

FOR THE AREA OF A THREE-LEVEL CROSS-SECTION.

c m n d f c, figs. 11 and 12.

$$\text{Area} = \frac{ab \times fe}{2} + \frac{mn(ac+bd)}{4}$$

SHRINKAGE OF EMBANKMENT.

Although earth, when first dug, and loosely thrown out, *swells* about $\frac{1}{5}$ part, so that a cubic yard *in place* averages about $1\frac{1}{5}$ or 1.2 cubic yards when dug; or 1 cubic yard dug is equal to $\frac{5}{6}$, or to .8333 of a cubic yard in place; yet when made into embankment it gradually subsides, settles, or shrinks, into a less bulk than it occupied *before being dug*.

The following are approximate averages of the shrinkage; or, in other words, the earth measured in place in a cut, will, when made into embankment, occupy a bulk less than before by about the following proportions:

Gravel or sand................about	8 per ct;	or 1 in 12½ less.
Clay.................................. "	10 per ct;	or 1 in 10 less.
Loam................................. "	12 per ct;	or 1 in 8½ less.
Loose vegetable surface soil.... "	15 per ct;	or 1 in 6¾ less.
Puddled clay....................... "	25 per ct;	or 1 in 4 less.

The writer thinks, from some trials of his own, that 1 cubic yard of any hard rock in place, will make from $1\frac{5}{8}$ to $1\frac{3}{4}$ cubic yards of embankment; say on an average 1.7 cubic yards. Or that 1 cubic yard of rock embankment requires .5882 of a cubic yard in place. He found that a solid cubic yard when broken into fragments, made about as follows:

	CUBIC YDS.	Of which there were	
		SOLID.	VOIDS.
In loose heap.........1.9		52.6 per cent.	47.4 per cent.
Carelessly piled......1.75		57 "	43 "
Carefully piled.......1.6		63 "	57 "

TABLE 1.—LEVEL CUTTINGS.

Roadway 14 feet wide, side-slopes 1½ to 1.

H'ght in ft.	0	.1	.2	.3	.4	.5	.6	.7	.8	.9
	cu. yds.	cu. yds.	cu. yds.	cu. yds.	cu. yds.	cu. yds.	cu. yds.	cu. yds	cu. yds.	cu. yds.
0		5.24	10.6	16.1	21.6	27.3	33.1	39.0	45.0	51.2
1	57.4	63.8	70.2	76.8	83.5	90.3	97.2	104.2	111.3	118.6
2	125.9	133.4	141.0	148.6	156.4	164.4	172.4	180.5	188.7	197.1
3	205.6	214.1	222.8	231.6	240.5	249.5	258.7	267.9	277.3	286.7
4	296.3	306.0	315.8	325.7	335.7	345.8	356.1	366.4	376.9	387.5
5	398.1	408.9	419.9	430.9	442.0	453.2	464.6	476.1	487.6	499.3
6	511.1	523.0	535.0	547.2	559.4	571.8	584.2	596.8	609.5	622.3
7	635.2	648.2	661.3	674.6	687.9	701.4	714.9	728.6	742.4	756.3
8	770.3	784.5	798.7	813.1	827.5	842.1	856.8	871.6	886.5	901.5
9	916.7	931.9	947.3	962.7	978.3	994.0	1010	1026	1042	1058
10	1074	1090	1107	1123	1140	1157	1174	1191	1208	1225
11	1243	1260	1278	1295	1313	1331	1349	1367	1385	1404
12	1422	1441	1459	1478	1497	1516	1535	1554	1574	1593
13	1613	1633	1652	1672	1692	1712	1733	1753	1773	1794
14	1815	1835	1856	1877	1898	1920	1941	1962	1984	2006
15	2028	2050	2072	2094	2116	2138	2161	2183	2206	2229
16	2252	2275	2298	2321	2344	2368	2391	2415	2439	2463
17	2487	2511	2535	2559	2584	2608	2633	2658	2683	2708
18	2733	2759	2784	2809	2835	2861	2886	2912	2938	2964
19	2991	3017	3044	3070	3097	3124	3151	3178	3205	3232
20	3259	3287	3314	3342	3370	3398	3426	3454	3482	3510
21	3539	3567	3596	3625	3654	3683	3712	3741	3771	3800
22	3830	3859	3889	3919	3949	3979	4009	4040	4070	4101
23	4132	4162	4193	4224	4255	4287	4318	4349	4381	4413
24	4444	4476	4508	4541	4573	4605	4638	4670	4703	4736
25	4769	4802	4835	4868	4901	4935	4968	5002	5036	5070
26	5104	5138	5172	5206	5241	5275	5310	5345	5380	5415
27	5450	5485	5521	5556	5592	5627	5663	5699	5735	5771
28	5807	5844	5880	5917	5953	5990	6027	6064	6101	6139
29	6176	6213	6251	6289	6326	6364	6402	6440	6479	6517
30	6556	6594	6633	6672	6711	6750	6789	6828	6867	6907
31	6946	6986	7026	7066	7106	7146	7186	7226	7267	7307
32	7348	7389	7430	7471	7512	7553	7595	7636	7678	7719
33	7761	7803	7845	7887	7929	7972	8014	8057	8099	8142
34	8185	8228	8271	8315	8358	8401	8445	8489	8532	8576
35	8620	8664	8709	8753	8798	8842	8887	8932	8976	9022
36	9067	9112	9157	9203	9248	9294	9340	9386	9432	9478
37	9524	9570	9617	9663	9710	9757	9804	9951	9898	9945
38	9993	10040	10088	10135	10183	10231	10279	10327	10375	10424
39	10472	10521	10569	10618	10667	10716	10765	10815	10864	10913
40	10963	11013	11062	11112	11162	11212	11263	11313	11364	11414
41	11465	11516	11567	11618	11669	11720	11771	11823	11874	11926
42	11978	12029	12081	12134	12186	12238	12291	12343	12396	12449
43	12502	12555	12608	12661	12715	12768	12822	12875	12929	12983
44	13037	13091	13145	13200	13254	13309	13363	13418	13473	13528
45	13583	13639	13694	13749	13805	13861	13916	13972	14028	14084
46	14141	14197	14254	14310	14367	14424	14480	14537	14595	14652
47	14709	14767	14824	14882	14940	14998	15056	15114	15172	15230
48	15289	15347	15406	15465	15524	15583	15642	15701	15761	15820
49	15880	15939	15999	16059	16119	16179	16239	16300	16360	16421
50	16481	16542	16603	16664	16725	16787	16848	16909	16971	17033
51	17094	17156	17218	17280	17343	17405	17467	17530	17593	17656
52	17719	17782	17845	17908	17971	18035	18098	18162	18226	18290
53	18354	18418	18482	18546	18611	18675	18740	18805	18870	18935
54	19000	19065	19131	19196	19262	19327	19393	19459	19525	19591
55	19657	19724	19790	19857	19923	19990	20057	20124	20191	20259
56	20326	20393	20461	20529	20596	20664	20732	20800	20869	20937
57	21005	21074	21143	21212	21280	21349	21419	21488	21557	21627
58	21696	21766	21836	21906	21976	22046	22116	22186	22257	22327
59	22398	22469	22540	22611	22682	22753	22825	22896	22968	23039
60	23111	23183	23255	23327	23399	23472	23544	23617	23689	23762

For continuation to 170 feet, see table 15.

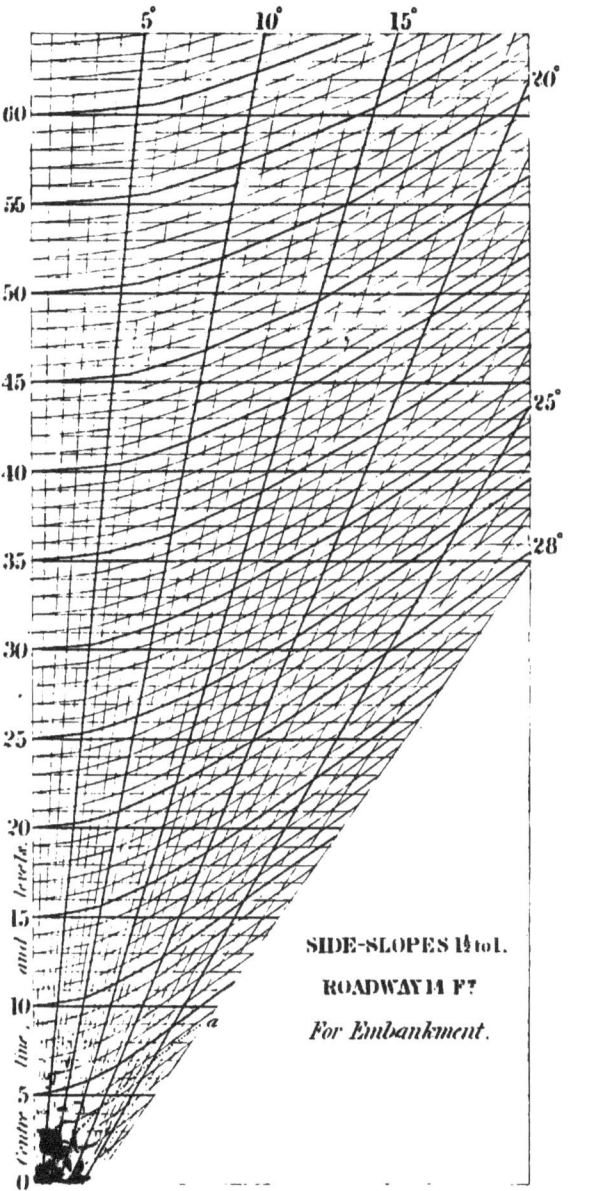

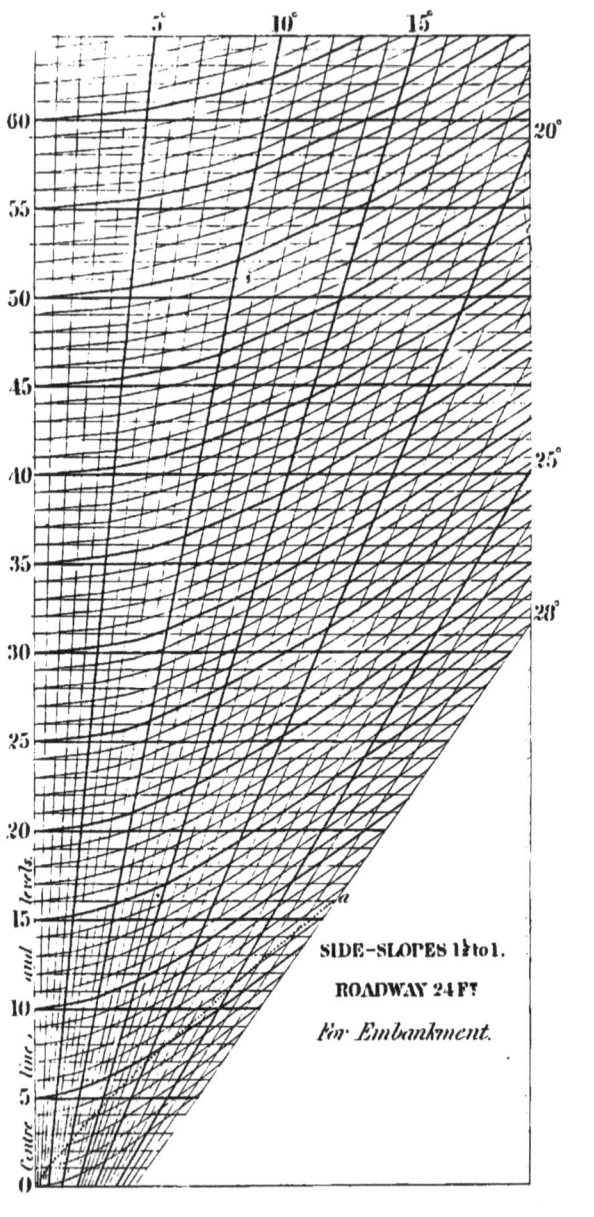

Plate 11.

TABLE 2.—LEVEL CUTTINGS.

Roadway 24 feet wide, side-slopes 1½ to 1.

H'ght in ft.	·0	·1	·2	·3	·4	·5	·6	·7	·8	·9
	cu. yds.	cu. yds.	cu. yds.	cu. yds.	cu. yds.	cu. yds.	cu. yds	cu. yds.	cu. yds.	cu. yds.
0		8·94	18·0	27·2	36·4	45·8	55·3	64·9	74·7	84·5
1	94·4	104·5	114·7	124·9	135·3	145·8	156·4	167·2	178·0	188·9
2	200·0	211·2	222·4	233·8	245·3	256·9	268·6	280·5	292·4	304·4
3	316·6	328·9	341·2	353·7	366·3	379·0	391·9	404·8	417·8	431·0
4	444·4	457·8	471·3	484·9	498·6	512·4	526·4	540·4	554·6	568·8
5	583·3	597·8	612·4	627·1	642·0	656·9	671·9	687·1	702·3	717·7
6	733·3	748·9	764·7	780·5	796·4	812·5	828·7	844·9	861·3	877·8
7	894·4	911·2	928·0	944·9	962·0	979·2	996·4	1014	1031	1049
8	1067	1085	1102	1121	1139	1157	1175	1194	1212	1231
9	1250	1269	1288	1307	1326	1346	1365	1385	1405	1425
10	1444	1465	1485	1505	1525	1546	1566	1587	1608	1629
11	1650	1671	1692	1714	1735	1757	1779	1800	1822	1845
12	1867	1889	1911	1934	1956	1979	2002	2025	2048	2071
13	2094	2118	2141	2165	2189	2213	2236	2261	2285	2309
14	2333	2358	2382	2407	2432	2457	2482	2507	2532	2558
15	2583	2609	2635	2661	2686	2713	2739	2765	2791	2818
16	2844	2871	2898	2925	2952	2979	3006	3034	3061	3089
17	3117	3145	3172	3201	3229	3257	3285	3314	3342	3371
18	3400	3429	3458	3487	3516	3546	3575	3605	3635	3665
19	3694	3725	3755	3785	3815	3846	3876	3907	3938	3969
20	4000	4031	4062	4094	4125	4157	4189	4221	4252	4285
21	4317	4349	4381	4414	4446	4479	4512	4545	4578	4611
22	4644	4678	4711	4745	4779	4813	4846	4881	4915	4940
23	4983	5018	5052	5087	5122	5157	5192	5227	5262	5298
24	5333	5369	5405	5441	5476	5513	5549	5585	5621	5658
25	5694	5731	5768	5805	5842	5879	5916	5954	5991	6029
26	6067	6105	6142	6191	6219	6257	6295	6334	6372	6411
27	6450	6489	6528	6567	6606	6646	6685	6725	6765	6805
28	6844	6985	6925	6965	7005	7046	7086	7127	7168	7209
29	7250	7291	7332	7374	7415	7457	7499	7541	7582	7625
30	7667	7709	7751	7794	7836	7879	7922	7965	8008	9051
31	8094	8138	8181	8225	8269	8313	8356	8401	8445	8489
32	8533	8578	8622	8607	8712	8757	8802	8847	8892	8938
33	8983	9029	9075	9121	9166	9212	9259	9305	9351	9398
34	9444	9491	9538	9585	9632	9679	9726	9774	9821	9869
35	9917	9965	10012	10061	10109	10157	10205	10254	10302	10351
36	10400	10449	10498	10547	10596	10646	10695	10745	10795	10845
37	10894	10945	10995	11045	11095	11146	11196	11247	11298	11349
38	11400	11451	11502	11554	11605	11657	11709	11761	11812	11865
39	11917	11969	12021	12074	12126	12179	12232	12285	12338	12391
40	12444	12498	12551	12605	12659	12713	12766	12821	12875	12929
41	12983	13038	13092	13147	13202	13257	13312	13367	13422	13478
42	13533	13589	13645	13701	13756	13813	13869	13925	13981	14038
43	14094	14151	14209	14265	14322	14379	14436	14494	14551	14609
44	14667	14725	14782	14840	14899	14957	15015	15074	15132	15191
45	15250	15309	15368	15427	15486	15546	15605	15665	15725	15785
46	15844	15905	15965	16025	16085	16146	16206	16267	16328	16389
47	16450	16511	16572	16634	16695	16757	16819	16881	16942	17005
48	17067	17129	17191	17254	17316	17379	17442	17505	17568	17631
49	17694	17758	17821	17885	17949	18013	18076	19141	18205	18269
50	18333	18398	18462	18527	18592	18657	18722	18787	18852	18918
51	18983	19049	19115	19181	19246	19313	19379	19445	19511	19578
52	19644	19711	19778	19845	19912	19979	20046	20114	20181	20249
53	20317	20385	20452	20521	20589	20657	20725	20794	20862	20931
54	21000	21069	21138	21207	21276	21346	21415	21485	21555	21625
55	21694	21765	21835	21905	21975	22046	22116	22187	22258	22329
56	22400	22471	22542	22614	22685	22757	22829	22901	22972	23045
57	23117	23189	23261	23334	23406	23479	23552	23625	23698	23771
58	23844	23918	23991	24065	24139	24213	24286	24361	24435	24509
59	24583	24658	24732	24807	24882	24957	25032	25107	25182	25258
60	25333	25409	25485	25561	25636	25713	25789	25865	25941	26018

For continuation to 170 feet, see table 16.

TABLE 3.—LEVEL CUTTINGS.
Roadway 18 feet wide, side-slopes 1 to 1.

Dep h in ft.	·0	·1	·2	·3	·4	·5	·6	·7	·8	·9
	cu. yds.	cu. yds.	cu. yds.	cu. yds.	cu. yds.	cu. yds.	cu. yds.	cu. yds.	cu. yds.	cu. yds.
0		6·7	13·5	20·3	27·3	34·3	41·3	48·5	55·7	63·0
1	70.4	77·8	85·3	92·9	100·6	108·3	116·1	124·0	132·0	140·0
2	148·1	156·3	164·6	172·9	181·3	189·8	198·4	207·0	215·7	224·5
3	233·3	242·3	251·3	260·3	269·5	278·7	288·0	297·4	306·8	316·3
4	325·9	335·6	345·3	355·1	365·0	375·0	385·0	395·1	405·3	415·6
5	425·9	436·3	446·8	457·4	468·0	478·7	489·5	500.3	511·3	522·3
6	533·3	544·5	555·7	567·0	578·4	589·8	601·3	612·9	624·6	636·5
7	646·1	680·0	672·0	684·0	696·1	708·3	720·6	732·9	745·3	757·8
8	770·4	783·0	795·7	808·5	821·3	834·3	847·3	860·3	873·5	886·7
9	900·0	913·4	926·8	940·3	953·9	967·6	981·3	995·1	1009	1023
10	1037	1051	1065	1080	1094	1108	1123	1137	1152	1167
11	1181	1196	1211	1226	1241	1256	1272	1287	1302	1318
12	1333	1349	1365	1380	1396	1412	1428	1444	1460	1476
13	1493	1509	1525	1542	1558	1575	1592	1608	1625	1642
14	1659	1676	1693	1711	1728	1745	1763	1780	1798	1816
15	1833	1851	1869	1887	1905	1923	1941	1960	1978	1996
16	2015	2033	2052	2071	2089	2108	2127	2146	2165	2184
17	2204	2223	2242	2262	2281	2301	2321	2340	2360	2380
18	2400	2420	2440	2460	2481	2501	2521	2542	2562	2583
19	2604	2624	2645	2666	2687	2708	2729	2751	2772	2793
20	2815	2836	2858	2880	2901	2923	2945	2967	2989	3011
21	3033	3056	3078	3100	3123	3145	3168	3191	3213	3236
22	3259	3282	3305	3328	3352	3375	3398	3422	3445	3469
23	3493	3516	3540	3564	3588	3612	3636	3660	3685	3709
24	3733	3758	3782	3807	3832	3856	3881	3906	3931	3956
25	3981	4007	4032	4057	4083	4108	4134	4160	4185	4211
26	4237	4263	4289	4315	4341	4368	4394	4420	4447	4473
27	4500	4527	4553	4580	4607	4634	4661	4688	4716	4743
28	4770	4798	4825	4853	4881	4908	4936	4964	4992	5020
29	5048	5076	5105	5133	5161	5190	5218	5247	5276	5304
30	5333	5362	5391	5420	5449	5479	5508	5537	5567	5596
31	5626	5656	5685	5715	5745	5775	5805	5835	5865	5896
32	5926	5956	5987	6017	6048	6079	6109	6140	6171	6202
33	6233	6264	6296	6327	6358	6390	6421	6453	6485	6516
34	6548	6580	6612	6644	6676	6708	6741	6773	6805	6838
35	6870	6903	6936	6968	7001	7034	7067	7100	7133	7167
36	7200	7233	7267	7300	7334	7368	7401	7435	7469	7503
37	7537	7571	7605	7640	7674	7708	7743	7777	7812	7847
38	7881	7916	7951	7986	8021	8056	8092	8127	8162	8198
39	8233	8269	8305	8340	8376	8412	8448	8484	8520	8556
40	8593	8629	8665	8702	8738	8775	8812	8848	8885	8922
41	8959	8996	9033	9071	9108	9145	9183	9220	9258	9296
42	9333	9371	9409	9447	9485	9523	9561	9600	9638	9676
43	9715	9753	9792	9831	9869	9908	9947	9986	10025	10064
44	10104	10143	10182	10222	10261	10301	10341	10380	10420	10460
45	10500	10540	10580	10620	10661	10701	10741	10782	10822	10863
46	10904	10944	10985	11026	11067	11108	11149	11191	11232	11273
47	11315	11356	11398	11440	11481	11523	11565	11607	11649	11691
48	11733	11776	11818	11860	11903	11945	11988	12031	12073	12116
49	12159	12202	12245	12288	12332	12375	12418	12462	12505	12549
50	12593	12636	12680	12724	12768	12812	12856	12900	12945	12989
51	13033	13078	13122	13167	13212	13256	13301	13346	13391	13436
52	13481	13527	13572	13617	13663	13708	13754	13800	13845	13891
53	13937	13983	14029	14075	14121	14168	14214	14260	14307	14353
54	14400	14447	14493	14540	14587	14634	14681	14728	14776	14823
55	14870	14918	14965	15013	15061	15108	15156	15204	15252	15300
56	15348	15396	15445	15493	15541	15590	15639	15687	15736	15784
57	15833	15882	15931	15980	16029	16079	16128	16177	16227	16276
58	16326	16376	16425	16475	16525	16575	16625	16675	16725	16776
59	16826	16876	16927	16977	17028	17079	17129	17180	17231	17282
60	17333	17384	17436	17487	17538	17590	17641	17693	17745	17796

For continuation to 170 feet, see table 15.

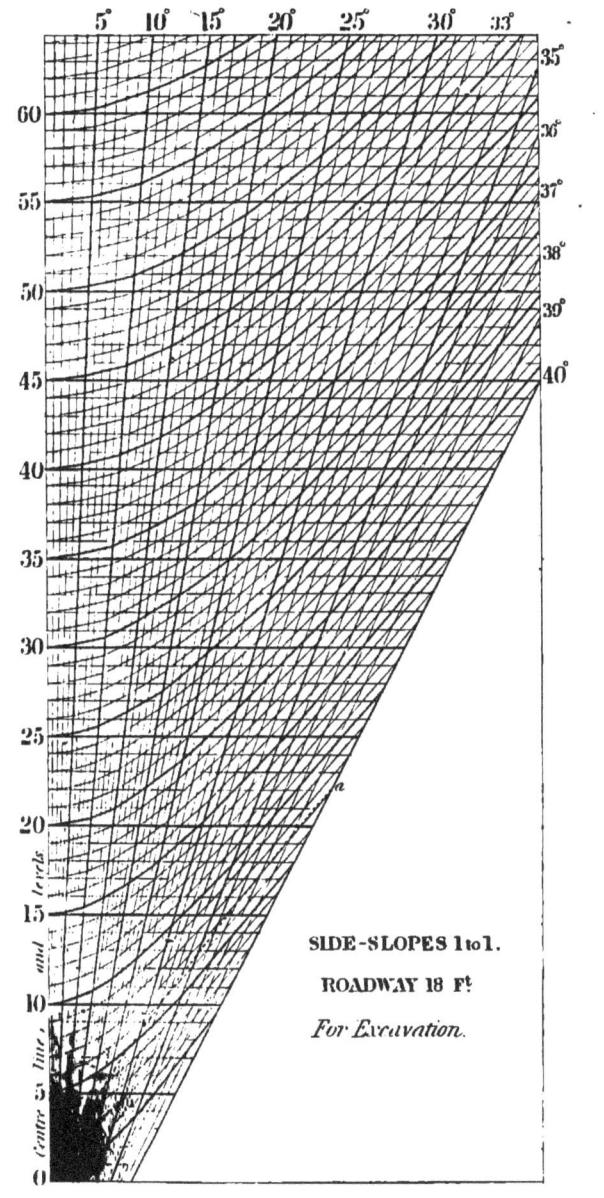

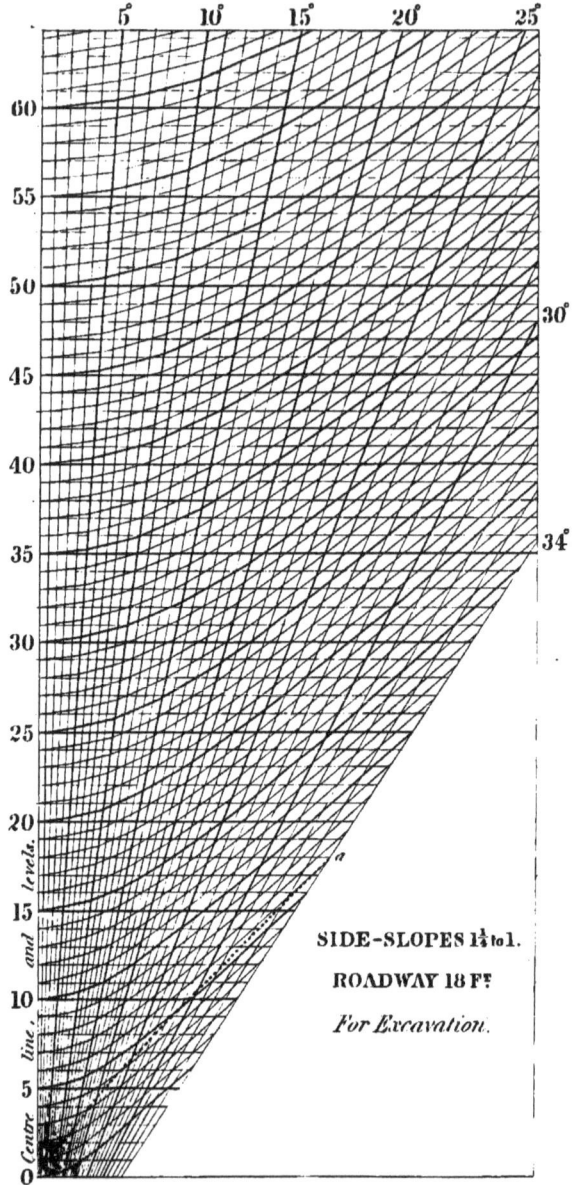

TABLE 4.—LEVEL CUTTINGS.
Roadway 18 feet wide, side-s'opes 1¼ to 1.

Depth in ft.	·0	·1	·2	·3	·4	·5	·6	·7	·8	9
	cu. yds.	cu. yds.	cu. yds.	cu. yds.	cu. yds.	cu. yds.	cu. yds.	cu. yds.	cu. yds.	cu. yds.
0		6·71	13·5	20·4	27·4	34·5	41·6	49·0	56·0	64·
1	71·3	78·9	86·7	94·5	102·4	110·4	118·5	126·7	135·0	143·
2	151·8	160·4	169·1	177·8	186·7	195·6	204·6	213·8	223·0	232·2
3	241·6	251·1	260·7	270·4	280·2	290·0	300·0	310·1	320·2	330·ξ
4	340·7	351·2	361·7	372·3	383·0	393·8	404·7	415·7	426·8	437·ϛ
5	449·2	460·6	472·1	483·7	495·3	507·0	518·8	530·8	542·8	554·ς
6	566·7	579·0	591·3	603·8	616·4	629·0	641·8	654·5	667·5	680·
7	693·5	706·7	720·0	733·4	746·9	760·4	774·1	787·8	801·7	815·(
8	829·6	843·8	858·0	872·3	886·7	901·2	915·8	930·4	945·2	960·1
9	975·0	990·0	1005	1020	1036	1051	1067	1082	1098	1114
10	1130	1146	1162	1178	1194	1210	1227	1243	1260	1277
11	1293	1310	1327	1344	1361	1379	1396	1413	1431	1448
12	1467	1484	1502	1520	1539	1557	1575	1593	1612	1630
13	1649	1668	1687	1706	1725	1744	1763	1782	1802	1821
14	1841	1860	1880	1900	1920	1940	1960	1980	2001	2021
15	2042	2062	2083	2104	2125	2146	2167	2188	2209	2231
16	2252	2273	2295	2317	2339	2361	2383	2405	2427	2449
17	2471	2494	2516	2539	2562	2585	2608	2631	2654	2677
18	2700	2723	2747	2770	2794	2818	2842	2866	2890	2914
19	2938	2962	2987	3011	3036	3060	3085	3110	3135	3160
20	3185	3210	3236	3261	3287	3312	3338	3364	3390	3416
21	3442	3468	3494	3520	3547	3573	3600	3627	3654	3680
22	3707	3734	3762	3789	3816	3844	3871	3899	3927	3954
23	3982	4010	4039	4067	4095	4123	4152	4180	4209	4238
24	4267	4296	4325	4354	4383	4412	4442	4471	4501	4530
25	4560	4590	4620	4650	4680	4710	4741	4771	4802	4832
26	4863	4894	4925	4956	4987	5018	5049	5080	5112	5143
27	5175	5207	5239	5270	5302	5334	5367	5399	5431	5464
28	5496	5529	5562	5594	5627	5660	5693	5727	5760	5793
29	5827	5860	5894	5928	5962	5996	6030	6064	6098	6132
30	6167	6201	6236	6270	6305	6340	6375	6410	6445	6480
31	6516	6551	6587	6622	6658	6694	6730	6766	6802	6838
32	6874	6910	6947	6983	7020	7057	7093	7130	7167	7204
33	7242	7279	7316	7354	7391	7429	7467	7504	7542	7580
34	7618	7656	7695	7733	7772	7810	7849	7887	7926	7965
35	8005	8044	8083	8122	8162	8201	8241	8280	8320	8360
36	8400	8440	8480	8520	8561	8601	8642	8682	8723	8764
37	8805	8846	8887	8928	8969	9010	9052	9093	9135	9177
38	9219	9260	9302	9344	9387	9429	9471	9514	9556	9599
39	9642	9684	9727	9770	9813	9857	9900	9943	9987	10030
40	10074	10118	10162	10206	10250	10294	10338	10382	10427	10471
41	10516	10560	10605	10650	10695	10740	10785	10830	10876	10921
42	10967	11012	11058	11104	11150	11196	11242	11288	11334	11380
43	11427	11473	11520	11567	11614	11660	11707	11754	11802	11849
44	11896	11944	11991	12039	12087	12134	12182	12230	12278	12327
45	12375	12423	12472	12520	12569	12618	12667	12716	12765	12814
46	12863	12912	12962	13011	13061	13111	13160	13210	13260	13310
47	13360	13410	13461	13511	13562	13613	13662	13713	13764	13815
48	13867	13918	13969	14020	14072	14123	14175	14227	14279	14330
49	14382	14435	14487	14539	14591	14644	14696	14749	14802	14855
50	14907	14960	15014	15067	15120	15173	15227	15280	15334	15388
51	15442	15496	15550	15604	15658	15712	15767	15821	15876	15931
52	15985	16040	16095	16150	16205	16260	16316	16371	16427	16482
53	16538	16594	16650	16706	16761	16817	16874	16930	16987	17043
54	17100	17157	17214	17270	17328	17385	17441	17499	17556	17613
55	17671	17729	17787	17845	17903	17961	18019	18077	18135	18193
56	18252	18310	18369	18428	18487	18545	18604	18663	18723	18782
57	18842	18901	18961	19021	19080	19140	19200	19260	19320	19380
58	19441	19501	19562	19622	19683	19744	19804	19865	19926	19988
59	20049	20110	20172	20233	20295	20357	20418	20480	20542	20604
60	20667	20729	20791	20854	20916	20979	21041	21104	21167	21230

For continuation to 170 feet see table 15.

TABLE 5.—LEVEL CUTTINGS.
Roadway 18 feet wide, side-slopes 1½ to 1.

Depth in ft.	.0	.1	.2	.3	.4	.5	.6	.7	.8	.9
	cu. yds.	cu. yds.	cu. yds.	cu. yds.	cu. yds.	cu. yds.	cu. yds.	cu. yds.	cu. yds.	cu. yds.
0		6.72	13.6	20.5	27.6	34.7	42.0	49.4	56.9	64.5
1	72.2	80.1	88.0	96.1	104.2	112.5	120.9	129.4	138.0	146.7
2	155.5	164.5	173.5	182.7	191.9	201.3	210.8	220.4	230.1	240.0
3	249.9	260.0	270.1	280.4	290.8	301.3	311.9	322.6	333.4	344.5
4	355.5	366.7	378.0	389.4	400.9	412.5	424.2	436.0	448.0	460.0
5	472.2	484.5	496.9	509.4	522.0	534.7	547.6	560.5	573.6	586.7
6	600.0	613.4	626.9	640.5	654.2	668.1	682.0	696.1	710.2	724.5
7	738.9	753.4	768.0	782.7	797.6	812.5	827.6	842.7	858.0	873.4
8	888.9	904.5	920.2	936.1	952.0	968.1	984.2	1001	1017	1033
9	1050	1067	1084	1101	1118	1135	1152	1169	1187	1205
10	1222	1240	1258	1276	1294	1313	1331	1349	1368	1387
11	1406	1425	1444	1463	1482	1501	1521	1541	1560	1580
12	1600	1620	1640	1661	1681	1701	1722	1743	1764	1785
13	1806	1827	1848	1869	1891	1913	1934	1956	1978	2000
14	2022	2045	2067	2089	2112	2135	2158	2181	2204	2227
15	2250	2273	2297	2321	2344	2368	2392	2416	2440	2465
16	2489	2513	2538	2563	2598	2613	2638	2663	2688	2713
17	2739	2765	2790	2816	2842	2868	2894	2921	2947	2973
18	3000	3027	3054	3081	3108	3135	3162	3189	3217	3245
19	3272	3300	3328	3356	3384	3413	3441	3469	3498	3527
20	3556	3585	3614	3643	3672	3701	3731	3761	3790	3820
21	3850	3880	3910	3941	3971	4001	4032	4063	4094	4125
22	4156	4187	4218	4249	4281	4313	4344	4376	4408	4440
23	4472	4505	4537	4569	4602	4635	4668	4701	4734	4767
24	4800	4833	4867	4901	4934	4968	5002	5036	5070	5105
25	5139	5173	5208	5243	5278	5313	5348	5383	5418	5453
26	5489	5525	5560	5596	5632	5668	5704	5741	5777	5813
27	5850	5887	5924	5961	5998	6035	6072	6109	6147	6185
28	6222	6260	6298	6336	6374	6413	6451	6489	6528	6567
29	6606	6645	6684	6723	6762	6801	6841	6881	6920	6960
30	7000	7040	7080	7121	7161	7201	7242	7283	7324	7365
31	7406	7447	7488	7529	7571	7613	7654	7696	7738	7780
32	7822	7865	7907	7949	7992	8035	8078	8121	8164	8207
33	8250	8293	8337	8381	8424	8468	8512	8556	8600	8645
34	8689	8733	8778	8823	8868	8913	8958	9003	9048	9093
35	9139	9185	9230	9276	9322	9368	9414	9461	9507	9553
36	9600	9647	9694	9741	9788	9835	9882	9929	9977	10025
37	10072	10120	10168	10216	10264	10313	10361	10409	10458	10507
38	10556	10605	10654	10703	10752	10801	10851	10901	10950	11000
39	11050	11100	11150	11200	11251	11301	11352	11403	11454	11505
40	11556	11607	11658	11709	11761	11813	11864	11916	11968	12020
41	12072	12125	12177	12229	12282	12335	12388	12441	12494	12547
42	12600	12653	12707	12761	12814	12868	12922	12976	13030	13085
43	13139	13193	13248	13303	13358	13413	13468	13523	13579	13633
44	13689	13745	13800	13856	13912	13968	14024	14081	14137	14193
45	14250	14307	14364	14421	14478	14535	14592	14649	14707	14765
46	14822	14880	14938	14996	15054	15113	15171	15229	15288	15347
47	15406	15465	15524	15583	15642	15701	15761	15821	15880	15940
48	16000	16060	16120	16181	16241	16301	16362	16423	16484	16545
49	16606	16667	16728	16789	16851	16913	16974	17036	17098	17160
50	17222	17285	17347	17409	17472	17535	17598	17661	17724	17787
51	17850	17913	17977	18041	18105	18168	18232	18296	18360	18425
52	18489	18553	18618	18683	18748	18813	18879	18943	19008	19073
53	19139	19205	19270	19336	19402	19468	19534	19601	19667	19733
54	19800	19867	19934	20000	20068	20135	20202	20269	20337	20405
55	20472	20540	20608	20676	20744	20813	20881	20949	21018	21087
56	21156	21225	21294	21363	21432	21501	21571	21641	21710	21780
57	21850	21920	21990	22061	22131	22201	22272	22343	22414	22485
58	22556	22627	22698	22769	22841	22913	22984	23056	23128	23200
59	23272	23345	23417	23489	23562	23635	23708	23781	23854	23927
60	24000	24073	24147	24221	24294	24368	24442	24516	24590	24665

For continuation to 170 feet see table 15.

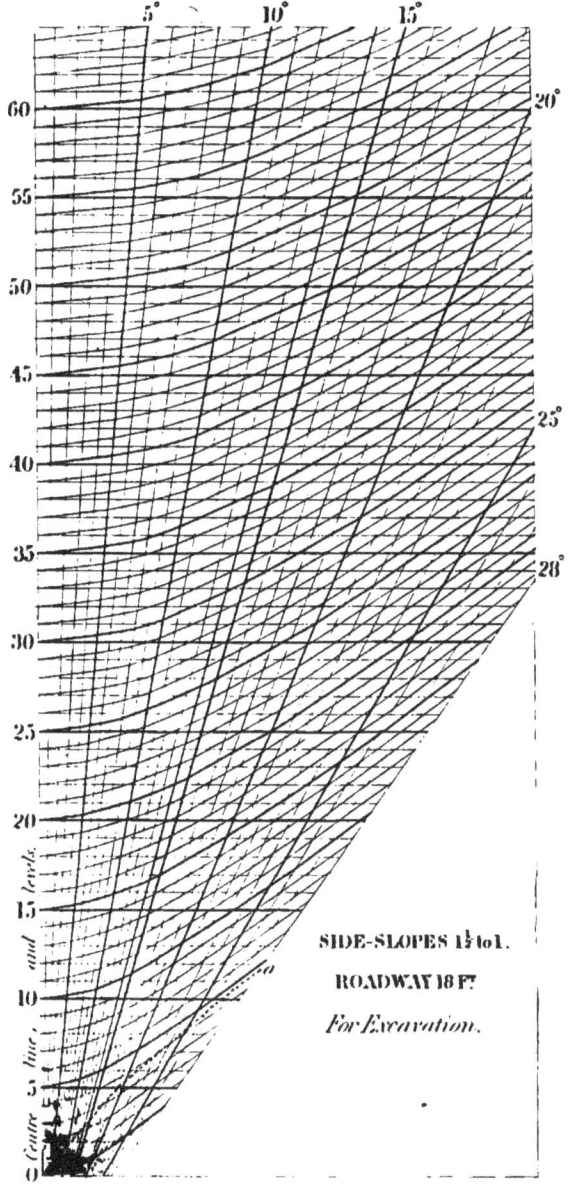

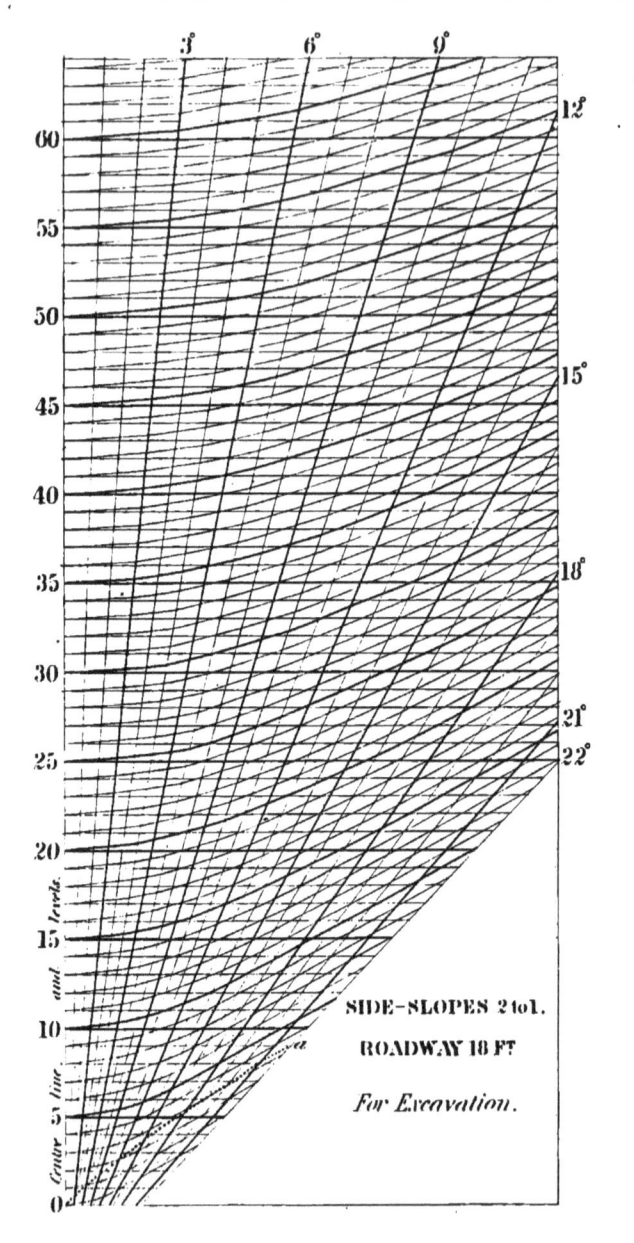

TABLE 6.—LEVEL CUTTINGS.

Roadway 18 feet wide, side-slopes 2 to 1.

Depth in ft.	·0	·1	·2	·3	·4	·5	·6	·7	·8	·9
	cu. yds.	cu. yds.	cu. yds.	u. yds.	u. yds.	cu. yds.	u. yds.	cu. yds.	cu. yds.	cu. yds.
0		6·74	13·6	20·7	27·9	35·2	42·7	50·3	58·1	66·0
1	74·1	82·3	90·7	99·2	107·9	116·7	125·6	134·7	144·0	153·4
2	163·0	172·7	182·5	192·5	202·7	213·0	223·4	234·0	244·7	255·6
3	266·7	277·9	289·2	300·7	312·3	324·1	336·0	348·1	360·3	372·7
4	385·2	397·9	410·7	423·6	436·7	450·0	463·4	477·0	490.7	504.5
5	519·5	532·7	547·0	561·4	576·0	590·7	605·6	620·7	635·9	651·2
6	666·7	682·3	698·1	714·0	730·1	746·3	762·7	779·2	795·9	812·7
7	829·6	846·7	864·0	881·4	899·0	916·7	934·5	952·5	970·7	989·0
8	1007	1026	1045	1064	1083	1102	1121	1141	1160	1190
9	1200	1220	1240	1261	1281	1302	1323	1344	1365	1386
10	1407	1429	1451	1473	1495	1517	1539	1561	1584	1607
11	1630	1653	1676	1699	1723	1746	1770	1794	1818	1842
12	1867	1891	1916	1941	1966	1991	2016	2041	2067	2093
13	2119	2145	2171	2197	2223	2250	2277	2304	2331	2358
14	2385	2413	2440	2468	2496	2524	2552	2581	2609	2638
15	2667	2696	2725	2754	2783	2813	2843	2873	2903	2933
16	2963	2993	3024	3055	3086	3117	3148	3179	3211	3242
17	3274	3306	3338	3370	3403	3435	3468	3501	3534	3567
18	3600	3633	3667	3701	3735	3769	3803	3837	3871	3906
19	3941	3976	4011	4046	4081	4117	4152	4188	4224	4260
20	4296	4333	4369	4406	4443	4480	4517	4554	4591	4629
21	4667	4705	4743	4781	4819	4857	4896	4935	4974	5013
22	5052	5091	5131	5170	5210	5250	5290	5330	5371	5411
23	5452	5493	5534	5575	5616	5657	5699	5741	5783	5825
24	5967	5909	5951	5994	6037	6080	6123	6166	6209	6253
25	6296	6340	6394	6428	6472	6517	6561	6606	6651	6696
26	6741	6786	6831	6877	6923	6969	7015	7061	7107	7153
27	7200	7247	7294	7341	7388	7435	7483	7530	7578	7626
28	7674	7722	7771	7819	7868	7917	7966	8015	8064	8113
29	8163	8213	8263	8313	8363	8413	8463	8514	8565	8616
30	8667	8718	8769	8821	8872	8924	8976	9028	9080	9133
31	9185	9238	9291	9344	9397	9450	9503	9557	9611	9665
32	9719	9773	9827	9881	9936	9991	10046	10101	10156	10211
33	10267	10322	10378	10434	10490	10546	10603	10659	10716	10773
34	10830	10887	10944	11001	11059	11117	11175	11233	11291	11349
35	11407	11466	11525	11584	11643	11702	11761	11821	11880	11940
36	12000	12060	12120	12181	12241	12302	12363	12424	12485	12546
37	12607	12669	12731	12793	12855	12917	12979	13041	13104	13167
38	13230	13293	13356	13419	13482	13546	13610	13674	13738	13802
39	13867	13931	13996	14061	14126	14191	14256	14321	14387	14453
40	14519	14585	14651	14717	14783	14850	14917	14984	15051	15118
41	15185	15253	15320	15388	15456	15524	15592	15661	15729	15798
42	15867	15936	16005	16074	16143	16213	16283	16353	16423	16493
43	16563	16633	16704	16775	16846	16917	16998	17059	17131	17202
44	17274	17346	17418	17490	17563	17635	17708	17781	17854	17927
45	18000	18073	18147	18221	18295	18369	18443	18517	18591	18666
46	18741	18816	18891	18966	19041	19117	19192	19268	19344	19420
47	19496	19573	19649	19726	19803	19880	19957	20034	20111	20189
48	20267	20345	20423	20501	20579	20657	20736	20815	20894	20973
49	21052	21131	21211	21290	21370	21450	21530	21610	21691	21771
50	21852	21933	22014	22095	22176	22257	22339	22421	22503	22585
51	22667	22749	22831	22914	22997	23080	23163	23246	23329	23413
52	23496	23580	23664	23748	23832	23917	24001	24086	24171	24256
53	24341	24426	24511	24597	24683	24769	24855	24941	25027	25113
54	25200	25287	25374	25461	25548	25635	25723	25810	25898	25986
55	26074	26162	26251	26339	26428	26517	26606	26695	26784	26873
56	26963	27053	27143	27233	27323	27413	27503	27594	27685	27776
57	27867	27958	28049	28141	28232	28324	28416	28508	28600	28693
58	28785	28878	28971	29064	29157	29250	29343	29437	29531	29625
59	29719	29813	29907	30001	30096	30191	30286	30381	30476	30571
60	30667	30762	30858	30954	31050	31146	31242	31339	31436	31533

For continuation to 170 feet, see table 15.

Table 7.—LEVEL CUTTINGS.
Roadway 28 feet wide, side-slopes 1 to 1.

Depth in ft.	·0	·1	·2	·3	·4	·5	·6	·7	·8	·9
	cu. yds	cu. yds.	cu. yds.	cu. yds.	cu. yds.	cu. yds.	cu. yds.	cu. yds.	cu. yds.	cu. yds.
0		10·4	20·9	31·4	42·1	52·8	63·6	74·4	85·3	96·3
1	107·4	118·6	129·8	141·1	152·4	163.9	175·4	187·0	198·7	210·4
2	222·2	234·1	246·1	258·1	270·2	282·4	294.7	307·0	319·4	331·9
3	344·4	357·1	369·8	382·6	395·4	408·3	421·3	434·4	447·6	460·8
4	474·1	487·4	500·9	514·4	528·0	541·7	555·4	569·2	583·1	597·1
5	611·1	625·2	639·4	653·7	668·0	682·4	696·9	711·4	726·1	740·8
6	755·6	770·4	785·4	800·4	815·5	830·6	945·8	861·1	876·5	891·9
7	907·5	923·0	938·7	954·5	970·3	986·2	1002	1018	1034	1050
8	1067	1083	1099	1116	1132	1149	1166	1182	1199	1216
9	1233	1250	1267	1285	1302	1319	1337	1354	1372	1390
10	1407	1425	1443	1461	1479	1497	1515	1534	1552	1570
11	1589	1607	1626	1645	1664	1682	1701	1720	1739	1759
12	1778	1797	1816	1836	1855	1875	1895	1914	1934	1954
13	1974	1994	2014	2034	2055	2075	2095	2116	2136	2157
14	2178	2199	2219	2240	2261	2282	2304	2325	2346	2367
15	2389	2410	2432	2454	2475	2497	2519	2541	2563	2585
16	2607	2630	2652	2674	2697	2719	2742	2765	2788	2810
17	2833	2856	2879	2903	2926	2949	2972	2996	3019	3043
18	3067	3090	3114	3138	3162	3186	3210	3234	3259	3283
19	3307	3332	3356	3381	3406	3431	3455	3480	3505	3530
20	3556	3581	3606	3631	3657	3682	3708	3734	3759	3785
21	3811	3837	3863	3889	3915	3942	3968	3994	4021	4047
22	4074	4101	4128	4154	4181	4208	4235	4263	4290	4317
23	4344	4372	4399	4427	4455	4482	4510	4538	4566	4594
24	4622	4650	4679	4707	4735	4764	4792	4821	4850	4879
25	4907	4936	4965	4994	5024	5053	5082	5111	5141	5170
26	5200	5230	5259	5289	5319	5349	5379	5409	5439	5470
27	5500	5530	5561	5591	5622	5653	5684	5714	5745	5776
28	5807	5839	5870	5901	5932	5964	5995	6027	6059	6090
29	6122	6154	6186	6218	6250	6292	6315	6347	6379	6412
30	6444	6477	6510	6543	6575	6608	6641	6674	6708	6741
31	6774	6807	6841	6874	6908	6942	6975	7009	7043	7077
32	7111	7145	7179	7214	7248	7282	7317	7351	7386	7421
33	7456	7490	7525	7560	7595	7631	7666	7701	7736	7772
34	7807	7843	7879	7914	7950	7996	8022	8058	8094	8130
35	8167	8203	8239	8276	8312	8349	8386	8423	8459	8496
36	8533	8570	8605	8645	8682	8719	8757	8794	8832	8870
37	8907	8945	8983	9021	9059	9097	9135	9174	9212	9250
38	9289	9327	9366	9405	9444	9482	9521	9560	9599	9639
39	9678	9717	9756	9796	9835	9875	9915	9954	9994	10034
40	10074	10114	10154	10194	10235	10275	10315	10356	10396	10437
41	10478	10519	10559	10600	10641	10682	10724	10765	10806	10847
42	10889	10930	10972	11014	11055	11097	11139	11181	11223	11265
43	11307	11350	11392	11434	11477	11519	11562	11605	11648	11690
44	11733	11776	11819	11863	11906	11949	11992	12036	12079	12123
45	12167	12210	12254	12298	12342	12386	12430	12474	12519	12563
46	12607	12652	12696	12741	12786	12831	12875	12920	12965	13010
47	13056	13101	13146	13191	13237	13282	13328	13374	13419	13465
48	13511	13557	13603	13649	13695	13742	13788	13834	13881	13927
49	13974	14021	14068	14114	14161	14208	14255	14303	14350	14397
50	14444	14492	14539	14587	14635	14682	14730	14778	14826	14874
51	14922	14970	15019	15067	15115	15164	15212	15261	15310	15359
52	15407	15456	15505	15554	15604	15653	15702	15751	15801	15850
53	15900	15950	15999	16049	16099	16149	16199	16249	16299	16350
54	16400	16450	16501	16551	16602	16653	16704	16754	16805	16856
55	16907	16959	17010	17061	17112	17164	17215	17267	17319	17370
56	17422	17474	17526	17578	17630	17682	17735	17787	17839	17892
57	17944	17997	18050	18103	18155	18208	18261	18314	18368	18421
58	18474	18527	18581	18634	18688	18742	18795	18849	18903	18957
59	19011	19065	19119	19174	19228	19282	19337	19391	19446	19501
60	19556	19610	19665	19720	19775	19831	19886	19941	19996	20052

For continuation to 170 feet see table 15.

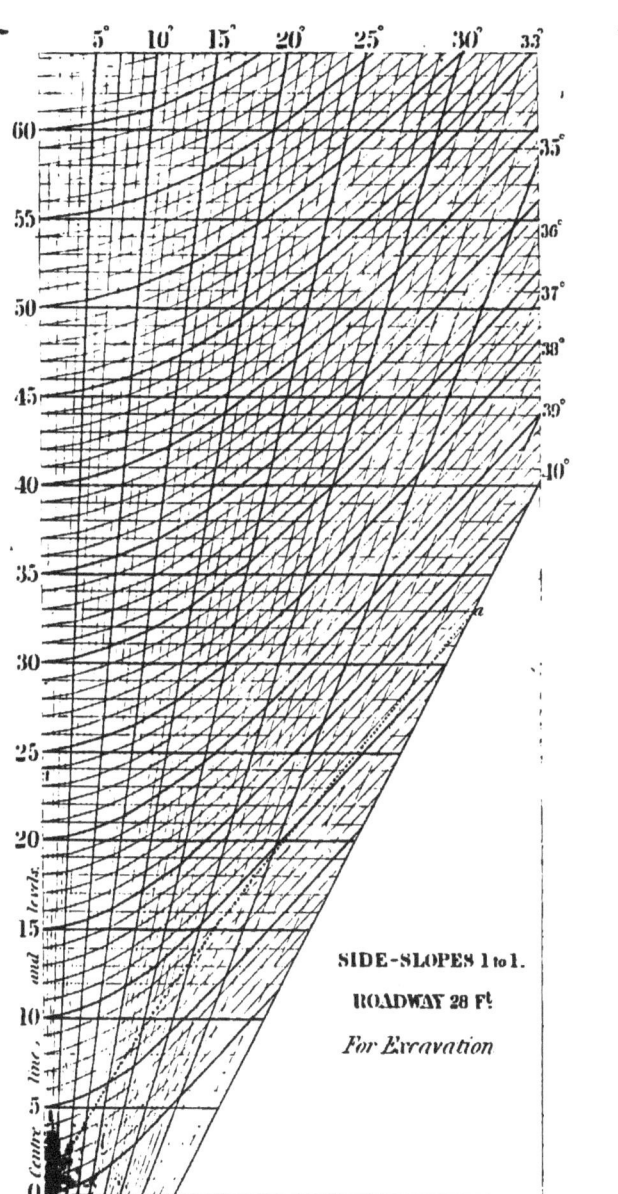

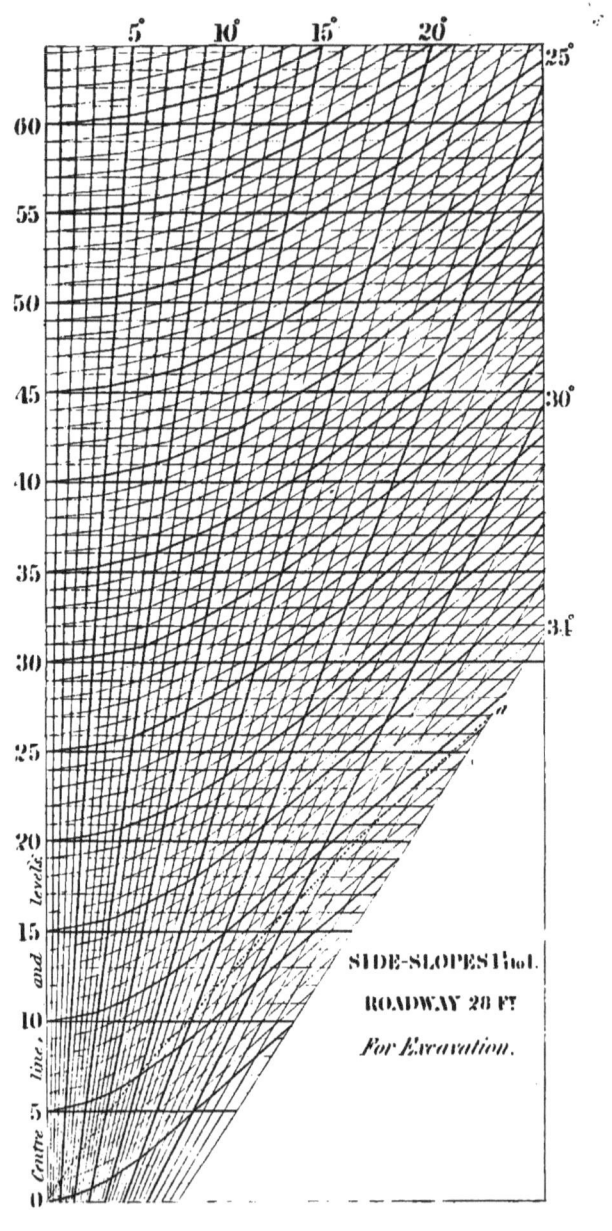

TABLE 8.—LEVEL CUTTINGS.
Roadway 28 feet wide, side-slopes 1¼ to 1.

Depth in ft.	0	·1	·2	·3	·4	·5	·6	·7	·8	·9
	cu. yds.	cu. yds.	cu. yds.	cu. yds.	cu. yds.	cu. yds.	cu. yds.	cu. yds.	cu. yds.	cu. yds.
0		10·4	20·9	31·5	42·2	53·0	63·9	74·9	85·9	97·1
1	108·3	119·7	131·1	142·7	154·3	166·0	177·8	189·7	201·7	213·8
2	225·9	238·2	250·6	263·0	275·6	288·2	301·0	313·8	326·7	339·8
3	352·8	366·0	379·3	392·7	406·2	419·8	433·4	447·2	461·1	475·0
4	488·9	503·0	517·3	531·6	546·0	560·5	575·1	589·8	604·6	619·5
5	634·4	649·5	664·7	679·9	695.3	710·7	726·2	741·9	757·6	773·4
6	788·9	804·9	821·0	837·2	853·4	869·8	886·3	902·8	919·5	936·2
7	952·8	969·8	986·8	1004	1021	1038	1056	1073	1091	1109
8	1126	1144	1162	1180	1198	1216	1234	1252	1271	1289
9	1308	1327	1346	1365	1384	1403	1422	1442	1461	1481
10	1500	1520	1540	1559	1579	1599	1620	1640	1660	1681
11	1701	1722	1742	1763	1784	1805	1826	1847	1869	1890
12	1911	1933	1954	1976	1998	2020	2042	2064	2086	2109
13	2131	2153	2175	2198	2221	2244	2267	2290	2313	2336
14	2359	2382	2406	2429	2453	2477	2501	2525	2549	2573
15	2597	2622	2646	2671	2695	2720	2745	2770	2795	2820
16	2944	2870	2895	2921	2946	2971	2998	3023	3049	3075
17	3101	3127	3153	3180	3206	3233	3259	3286	3313	3340
18	3367	3393	3421	3448	3475	3503	3531	3558	3596	3614
19	3641	3669	3698	3726	3754	3782	3811	3840	3868	3897
20	3925	3954	3984	4013	4042	4071	4101	4130	4160	4189
21	4219	4249	4279	4309	4339	4369	4399	4430	4460	4491
22	4522	4553	4584	4615	4646	4677	4708	4740	4771	4803
23	4834	4866	4898	4930	4962	4994	5026	5058	5091	5123
24	5156	5188	5221	5254	5287	5320	5353	5386	5419	5453
25	5486	5520	5553	5587	5621	5655	5689	5723	5757	5791
26	5826	5860	5895	5930	5964	5999	6034	6069	6104	6140
27	6175	6210	6246	6282	6317	6353	6389	6425	6461	6497
28	6533	6570	6606	6643	6679	6716	6753	6790	6827	6864
29	6901	6938	6976	7013	7051	7088	7126	7164	7202	7240
30	7278	7316	7354	7393	7431	7470	7508	7547	7586	7625
31	7664	7703	7742	7781	7821	7860	7900	7940	7979	8019
32	8059	8099	8139	8180	8220	8260	8301	8341	8382	8423
33	8464	8505	8546	8587	8628	8669	8711	8752	8794	8836
34	8878	8920	8962	9004	9046	9088	9131	9173	9216	9258
35	9301	9344	9387	9430	9473	9516	9559	9603	9646	9689
36	9733	9777	9821	9865	9909	9953	9997	10041	10086	10130
37	10175	10220	10264	10309	10354	10399	10445	10490	10535	10580
38	10626	10672	10717	10763	10809	10855	10901	10947	10993	11040
39	11086	11133	11179	11226	11273	11320	11367	11414	11461	11508
40	11556	11603	11651	11698	11746	11794	11842	11890	11938	11986
41	12035	12083	12131	12180	12228	12277	12326	12375	12424	12473
42	12522	12572	12621	12671	12720	12770	12819	12869	12919	12969
43	13019	13069	13120	13171	13221	13272	13322	13373	13424	13475
44	13526	13577	13628	13680	13731	13782	13834	13886	13938	13990
45	14042	14094	14146	14198	14251	14303	14356	14408	14461	14514
46	14567	14620	14673	14726	14779	14833	14886	14940	14993	15047
47	15101	15155	15209	15263	15317	15372	15426	15480	15535	15590
48	15644	15699	15754	15810	15864	15920	15975	16030	16086	16142
49	16197	16253	16309	16365	16421	16477	16533	16590	16646	16703
50	16759	16816	16873	16930	16987	17044	17101	17158	17216	17273
51	17331	17388	17446	17504	17562	17620	17678	17736	17794	17853
52	17911	17970	18028	18087	18146	18205	18264	18323	18382	18442
53	18500	18560	18620	18680	18739	18799	18859	18919	18980	19040
54	19100	19160	19220	19282	19342	19403	19464	19525	19586	19647
55	19708	19770	19831	19893	19954	20016	20078	20140	20202	20264
56	20326	20388	20451	20513	20576	20639	20701	20764	20827	20890
57	20953	21016	21079	21143	21206	21270	21333	21397	21461	21525
58	21589	21653	21717	21782	21846	21911	21975	22040	22104	22169
59	22234	22299	22364	22430	22495	22561	22626	22692	22758	22823
60	22889	22955	23021	23087	23153	23220	23286	23353	23419	23486

For continuation to 170 feet see Table 15.

TABLE 9.—LEVEL CUTTINGS.

Roadway 28 feet wide, side-slopes 1½ to 1.

Depth in ft.	·0	·1	·2	·3	·4	·5	·6	·7	·8	·9
	cu. yds.	cu. yds.	cu. yds.	cu. yds.	cu. yds.	cu. yds.	cu. yds.	cu. yds.	cu. yds.	cu. yds.
0		10·4	21·0	31·6	42·4	53·2	64·2	75·3	86·5	97·9
1	109·3	120·8	132·5	144·3	156·1	168·1	180·2	192·4	204·8	217·2
2	229·6	242·3	255·0	267·9	280·9	294·0	307·2	320·5	334·0	347·5
3	361·2	374·9	388·8	402·8	416·9	431·1	445·4	459·9	474·4	489·1
4	503·7	518·6	533·6	548·6	563·9	579·3	594·7	610·2	625·8	641·6
5	657·5	673·4	689·5	705·7	722·1	738·5	755·0	771·7	788·4	805·3
6	822·2	839·3	856·5	873·8	891·2	908·8	926·4	944·2	962·0	980·0
7	998·1	1016	1035	1053	1072	1090	1109	1128	1147	1166
8	1185	1204	1224	1243	1263	1283	1303	1322	1343	1363
9	1383	1403	1424	1445	1465	1486	1507	1528	1549	1571
10	1592	1614	1635	1657	1679	1701	1723	1745	1767	1790
11	1812	1835	1858	1881	1904	1927	1950	1973	1997	2020
12	2044	2068	2092	2116	2140	2164	2189	2213	2238	2262
13	2287	2312	2337	2362	2387	2413	2438	2464	2489	2515
14	2541	2567	2593	2619	2645	2672	2698	2725	2752	2779
15	2806	2833	2860	2887	2915	2942	2970	2997	3025	3053
16	3081	3109	3138	3166	3195	3223	3252	3281	3310	3339
17	3368	3397	3427	3456	3486	3516	3546	3576	3606	3636
18	3667	3697	3728	3758	3789	3820	3851	3882	3913	3944
19	3976	4007	4039	4070	4102	4134	4166	4198	4231	4263
20	4296	4328	4361	4394	4427	4460	4493	4527	4560	4594
21	4627	4661	4695	4729	4763	4797	4832	4866	4900	4935
22	4970	5005	5040	5075	5111	5146	5181	5217	5253	5288
23	5324	5360	5396	5432	5469	5505	5542	5578	5615	5652
24	5689	5726	5763	5800	5838	5875	5913	5951	5989	6027
25	6065	6103	6141	6179	6218	6257	6295	6334	6373	6412
26	6451	6491	6530	6570	6609	6649	6689	6729	6769	6809
27	6850	6890	6931	6971	7012	7053	7094	7135	7176	7217
28	7259	7300	7342	7384	7426	7468	7510	7552	7594	7637
29	7680	7722	7765	7808	7851	7894	7937	7981	8024	8067
30	8111	8155	8199	8243	8287	8331	8375	8420	8464	8509
31	8554	8598	8643	8688	8734	8779	8824	8870	8915	8961
32	9007	9053	9099	9145	9191	9238	9284	9331	9378	9425
33	9472	9519	9566	9613	9661	9708	9756	9804	9851	9900
34	9948	9997	10045	10093	10142	10190	10239	10288	10337	10386
35	10435	10484	10534	10583	10633	10683	10732	10782	10832	10882
36	10933	10983	11034	11084	11135	11186	11237	11288	11339	11391
37	11443	11494	11546	11598	11649	11701	11753	11806	11858	11910
38	11963	12016	12068	12121	12174	12227	12281	12334	12387	12441
39	12494	12548	12602	12656	12710	12764	12819	12873	12928	12982
40	13037	13092	13147	13202	13257	13312	13368	13423	13479	13535
41	13591	13647	13703	13759	13815	13872	13928	13985	14042	14099
42	14156	14213	14270	14327	14385	14442	14500	14558	14615	14673
43	14731	14790	14848	14906	14965	15024	15082	15141	15200	15259
44	15318	15378	15437	15497	15556	15616	15676	15736	15796	15856
45	15917	15977	16038	16098	16159	16220	16281	16342	16403	16465
46	16526	16587	16649	16711	16773	16835	16897	16959	17021	17084
47	17146	17209	17272	17335	17398	17461	17524	17587	17651	17714
48	17778	17842	17905	17969	18033	18098	18162	18226	18291	18356
49	18420	18485	18550	18615	18680	18746	18811	18877	18942	19008
50	19074	19140	19206	19272	19339	19405	19472	19538	19605	19672
51	19739	19806	19873	19940	20008	20075	20143	20211	20279	20347
52	20415	20483	20551	20620	20688	20757	20826	20894	20963	21032
53	21102	21171	21241	21310	21380	21450	21519	21589	21659	21730
54	21800	21870	21941	22012	22082	22153	22224	22295	22366	22438
55	22509	22581	22652	22724	22796	22868	22940	23012	23085	23157
56	23230	23302	23375	23448	23521	23594	23667	23741	23814	23888
57	23961	24035	24109	24183	24257	24331	24405	24480	24554	24629
58	24704	24779	24854	24929	25004	25079	25155	25230	25306	25381
59	25457	25533	25609	25686	25762	25838	25915	25992	26068	26145
60	26222	26299	26376	26454	26531	26609	26686	26764	26842	26920

For continuation to 170 feet see table 15.

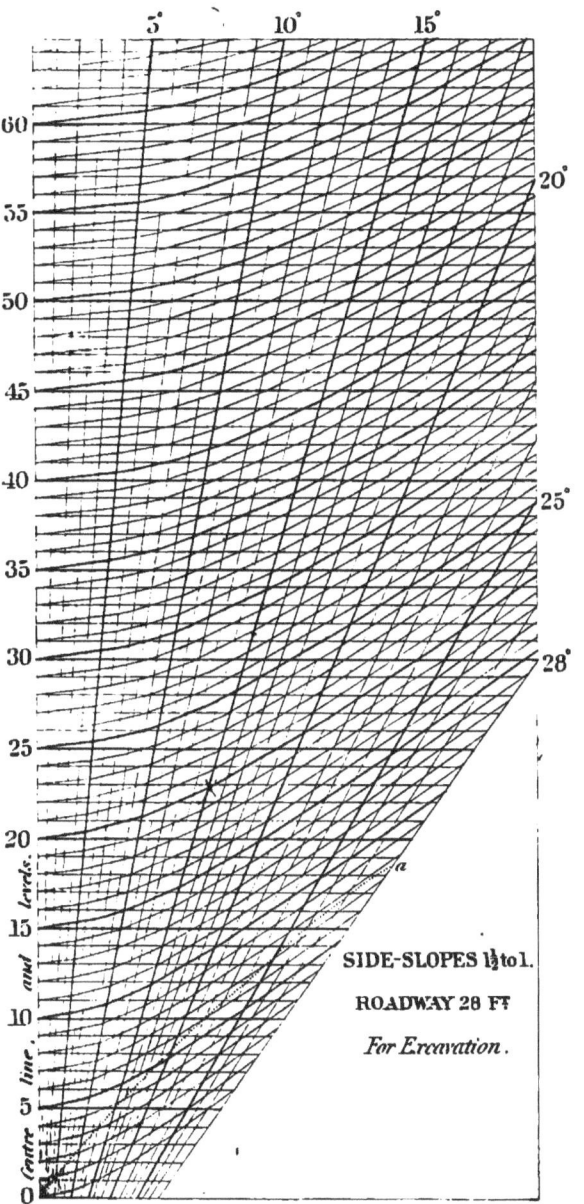

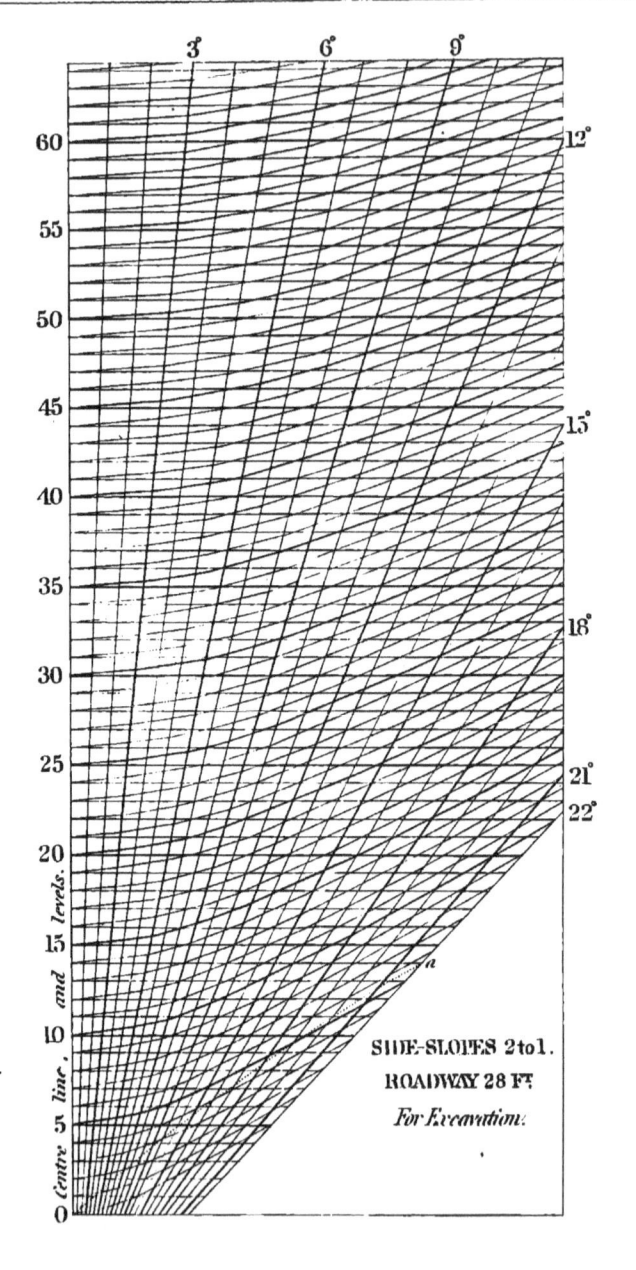

TABLE 10.—LEVEL CUTTINGS.

Roadway 28 feet wide, side-slopes 2 to 1.

Depth in ft.	·0	·1	·2	·3	·4	5.	·6	·7	·8	·9
	cu. yds.	cu. yds.	cu. yds.	cu. yds.	cu. yds	cu. yds.	cu. yds.	cu. yds.	cu. yds.	cu. yds.
0		10·4	21·0	31·8	42·7	53·7	64·9	76·2	87·7	99·3
1	111·1	123·0	135·1	147·3	159·7	172·2	184·9	197·7	210·7	223·8
2	237·0	250·4	264·0	277·7	291·6	305·6	319·7	334·0	348·4	363·0
3	377·8	392·7	407·7	422·9	438·2	453·7	469·3	485·1	501·0	517·1
4	533·3	549·7	566·2	582·9	599·7	616·6	633·7	651·0	668·4	686·0
5	703·7	721·5	739·5	757·7	776·0	794·4	813·0	831·7	850·6	869·6
6	888·9	908·2	927·7	948·3	968·1	988·1	1008	1028	1049	1069
7	1090	1111	1132	1153	1174	1195	1217	1239	1260	1282
8	1305	1327	1349	1372	1394	1417	1440	1464	1487	1510
9	1533	1557	1581	1605	1629	1654	1678	1703	1728	1753
10	1778	1803	1828	1854	1880	1905	1931	1958	1984	2010
11	2037	2064	2091	2118	2145	2172	2200	2227	2255	2283
12	2311	2339	2368	2396	2425	2454	2483	2512	2541	2570
13	2600	2630	2659	2689	2720	2750	2780	2811	2842	2873
14	2904	2935	2966	2998	3029	3061	3093	3125	3157	3190
15	3222	3255	3288	3321	3354	3387	3420	3454	3488	3521
16	3555	3590	3624	3658	3693	3728	3763	3798	3833	3868
17	3904	3939	3975	4011	4047	4083	4120	4156	4193	4230
18	4267	4304	4341	4378	4416	4454	4492	4529	4568	4606
19	4644	4683	4722	4761	4800	4839	4878	4918	4957	4997
20	5037	5077	5117	5158	5198	5239	5280	5321	5362	5403
21	5444	5486	5528	5570	5612	5654	5696	5738	5781	5824
22	5867	5910	5953	5996	6040	6083	6127	6171	6215	6259
23	6303	6348	6393	6438	6483	6528	6573	6618	6664	6710
24	6756	6802	6848	6894	6941	6987	7034	7081	7128	7175
25	7222	7270	7317	7365	7413	7461	7509	7558	7606	7655
26	7704	7753	7802	7851	7900	7950	8000	8049	8099	8150
27	8200	8250	8301	8352	8403	8454	8505	8556	8608	8659
28	8711	8763	8815	8867	8920	8972	9025	9078	9131	9184
29	9237	9290	9344	9398	9451	9506	9560	9614	9668	9723
30	9778	9833	9888	9943	9998	10054	10109	10165	10221	10277
31	10333	10390	10446	10503	10560	10617	10674	10731	10788	10846
32	10904	10961	11019	11078	11136	11194	11253	11312	11371	11430
33	11489	11548	11608	11667	11727	11787	11847	11907	11968	12028
34	12089	12150	12211	12272	12333	12394	12456	12518	12580	12642
35	12704	12766	12828	12991	12954	13017	13080	13143	13206	13270
36	13333	13397	13461	13525	13589	13654	13718	13783	13848	13913
37	13978	14043	14108	14174	14240	14305	14371	14438	14504	14570
38	14637	14704	14770	14838	14905	14972	15040	15107	15175	15243
39	15311	15379	15448	15516	15585	15654	15723	15792	15861	15930
40	16000	16070	16140	16210	16280	16350	16420	16491	16562	16633
41	16704	16775	16846	16918	16989	17061	17133	17205	17277	17350
42	17422	17495	17568	17641	17714	17787	17860	17934	18008	18081
43	18156	18230	18304	18378	18453	18528	18603	18678	18753	18828
44	18904	18979	19055	19131	19207	19283	19360	19436	19513	19590
45	19667	19744	19821	19898	19976	20054	20132	20210	20288	20366
46	20444	20523	20602	20681	20760	20839	20918	20998	21077	21157
47	21237	21317	21397	21478	21558	21639	21720	21801	21882	21963
48	22044	22126	22208	22290	22372	22454	22536	22618	22701	22784
49	22867	22950	23033	23116	23200	23283	23367	23451	23535	23619
50	23704	23788	23873	23958	24043	24128	24213	24298	24384	24470
51	24556	24642	24728	24814	24900	24987	25074	25161	25248	25335
52	25422	25510	25597	25685	25773	25861	25949	26038	26126	26215
53	26304	26393	26482	26571	26660	26750	26840	26930	27020	27110
54	27200	27290	27381	27472	27563	27654	27745	27836	27928	28019
55	28111	28203	28295	28387	28480	28572	28665	28758	28851	28944
56	29037	29130	29224	29318	29412	29506	29600	29694	29788	29883
57	29978	30073	30168	30263	30358	30454	30549	30645	30741	30837
58	30933	31030	31126	31223	31320	31417	31514	31611	31708	31806
59	31904	32002	32100	32199	32296	32394	32493	32592	32691	32790
60	32889	32988	33089	33187	33287	33387	33487	33587	33688	33788

For continuation to 170 feet see table 15.

TABLE 11.—LEVEL CUTTINGS.

Roadway 18 feet wide, side-slopes, ½ to 1.

Depth in ft.	·0	·1	·2	·3	·4	·5	·6	7.	·8	·9
	cu. yds.	cu. yds.	cu. yds.	cu. yds.	cu. yds.	cu. yds.	cu. yds.	cu. yds.	cu. yds.	cu. yds.
0		6·68	13·4	20·1	26·8	33·6	40·3	47·1	53·9	60·8
1	67·6	74·5	81·3	88·2	95·1	102·1	109·0	116·0	123·0	130·0
2	137·0	144·1	151·1	158·2	165·3	172·4	179·5	186·7	193·9	201·0
3	208·2	215·5	222·7	230·0	237·2	244·5	251·9	259·2	266·5	274·0
4	281·5	288·9	296·3	303·8	311·3	318·8	326·3	333·8	341·3	348·9
5	356·5	364·1	371·7	379·3	387·0	394·7	402·4	410·1	417·8	425·6
6	433·3	441·1	448·9	456·7	464·6	472·5	480·3	488·2	496·1	504·1
7	512·0	520·0	528·0	536·0	544·1	552·1	560·2	568·3	576·4	584·0
8	592·6	600·8	608·9	617·1	625·3	633·6	641·8	650·1	658·4	666·7
9	675·0	683·3	691·7	700·1	708·5	716·9	725·3	733·8	742·3	750·7
10	759·3	767·8	776·3	784·9	793·5	802·1	810·7	819·3	828·0	836·7
11	845·4	854·1	862·8	871·6	880·3	889·1	897·9	906·7	915·6	924·4
12	933·3	942·2	951·1	960·1	969·0	978·0	987·0	996·0	1005	1014
13	1023	1032	1041	1050	1060	1069	1078	1087	1096	1106
14	1115	1124	1133	1143	1152	1161	1171	1180	1189	1199
15	1208	1218	1227	1237	1246	1256	1265	1275	1284	1294
16	1304	1313	1323	1333	1342	1352	1362	1372	1381	1391
17	1401	1411	1421	1430	1440	1450	1460	1470	1480	1490
18	1500	1510	1520	1530	1540	1550	1560	1570	1580	1591
19	1601	1611	1621	1631	1642	1652	1662	1672	1683	1693
20	1703	1714	1724	1735	1745	1755	1766	1777	1787	1798
21	1808	1819	1829	1840	1850	1861	1872	1883	1894	1905
22	1916	1926	1937	1947	1958	1969	1980	1991	2002	2013
23	2024	2035	2046	2057	2068	2079	2090	2101	2111	2122
24	2133	2144	2156	2167	2178	2189	2200	2212	2223	2234
25	2245	2256	2268	2279	2290	2301	2313	2324	2336	2347
26	2359	2371	2382	2394	2406	2417	2429	2440	2452	2464
27	2475	2486	2498	2510	2521	2533	2545	2557	2569	2581
28	2592	2604	2616	2628	2640	2652	2664	2676	2688	2700
29	2712	2724	2736	2748	2760	2772	2784	2797	2809	2821
30	2833	2845	2858	2870	2882	2894	2907	2919	2931	2944
31	2956	2969	2981	2994	3006	3018	3031	3044	3056	3069
32	3081	3094	3107	3119	3132	3145	3157	3170	3183	3195
33	3208	3221	3234	3247	3260	3272	3285	3298	3311	3324
34	3337	3350	3363	3376	3389	3402	3415	3428	3441	3454
35	3467	3481	3494	3507	3520	3533	3547	3560	3573	3587
36	3600	3613	3627	3640	3653	3667	3680	3694	3707	3721
37	3734	3748	3761	3775	3788	3802	3816	3829	3843	3857
38	3870	3884	3898	3911	3925	3939	3953	3967	3980	3994
39	4008	4022	4036	4050	4064	4078	4092	4106	4120	4134
40	4148	4162	4176	4190	4205	4219	4233	4247	4261	4275
41	4290	4304	4318	4333	4347	4361	4376	4390	4404	4419
42	4433	4448	4462	4477	4491	4506	4520	4535	4549	4564
43	4579	4593	4608	4623	4637	4652	4667	4681	4696	4711
44	4726	4741	4756	4770	4785	4800	4815	4830	4845	4860
45	4875	4890	4905	4920	4935	4950	4965	4980	4995	5011
46	5026	5041	5056	5071	5087	5102	5117	5133	5148	5163
47	5179	5194	5209	5225	5240	5256	5271	5287	5302	5318
48	5333	5349	5364	5380	5396	5411	5427	5443	5458	5474
49	5490	5505	5521	5537	5553	5569	5585	5600	5616	5632
50	5648	5664	5680	5696	5712	5728	5744	5760	5776	5792
51	5808	5824	5840	5857	5873	5889	5905	5921	5938	5954
52	5970	5987	6003	6019	6036	6052	6068	6085	6101	6118
53	6134	6151	6167	6184	6200	6217	6233	6250	6267	6283
54	6300	6317	6333	6350	6367	6383	6400	6417	6434	6451
55	6468	6484	6501	6518	6535	6552	6569	6586	6603	6620
56	6637	6654	6671	6688	6705	6722	6739	6757	6774	6791
57	6808	6825	6843	6860	6877	6895	6912	6929	6947	6964
58	6981	6999	7016	7034	7051	7069	7086	7104	7121	7139
59	7156	7174	7192	7209	7227	7245	7262	7280	7298	7315
60	7333	7351	7369	7387	7405	7422	7440	7458	7476	7494

For continuation to 170 feet see Table 15.

Table 12.—LEVEL CUTTINGS.

Roadway 28 feet wide, side-slopes ½ to 1.

Depth in ft.	·0	·1	·2	·3	·4	·5	·6	·7	·8	·9
	cu. yds.	cu. yds.	cu. yds.	cu. yds.	cu. yds.	cu. yds	cu. yds	cu. yds.	cu. yds.	cu. yds.
0		10·4	20·8	31·2	41·6	52·1	62·6	73·0	83·6	94·1
1	104·6	115·2	125·8	136·4	147·0	157·6	168·3	179·0	189·7	200·4
2	211·1	221·9	232·6	243·4	254·3	265·1	275·9	286·8	297·7	308·6
3	319·4	330·4	341·3	352·3	363·3	374·3	385·4	396·4	407·5	418·6
4	429·7	440·8	451·9	463·1	474·3	485·5	496·7	508·0	519·2	530·5
5	541·7	553·0	564·3	575·6	587·0	598·4	609·7	621·1	632·6	644·0
6	655·6	667·1	678·7	690·3	702·0	713·6	725·3	737·0	747·7	759·4
7	771·3	783·0	794·7	806·4	818·1	829·9	841·6	853·4	865·3	877·1
8	888·9	900·8	912·6	923·6	935·5	947·4	959·4	971·4	983·4	995·4
9	1007	1019	1032	1044	1056	1068	1080	1092	1105	1117
10	1130	1142	1154	1166	1179	1191	1203	1216	1228	1240
11	1253	1265	1278	1290	1303	1315	1328	1340	1353	1365
12	1378	1390	1403	1416	1428	1441	1454	1466	1479	1492
13	1505	1517	1530	1543	1556	1569	1582	1595	1607	1620
14	1633	1646	1659	1672	1685	1699	1712	1725	1738	1751
15	1764	1777	1790	1803	1817	1830	1843	1856	1870	1883
16	1896	1910	1923	1936	1950	1963	1977	1990	2004	2017
17	2031	2044	2058	2071	2085	2098	2112	2126	2139	2153
18	2167	2180	2194	2208	2222	2235	2249	2263	2277	2291
19	2305	2318	2332	2346	2360	2374	2388	2402	2416	2430
20	2444	2459	2473	2487	2501	2515	2529	2543	2558	2572
21	2586	2600	2615	2629	2643	2658	2672	2686	2701	2715
22	2730	2744	2759	2773	2789	2802	2817	2831	2846	2860
23	2875	2890	2904	2919	2934	2948	2963	2978	2993	3007
24	3022	3037	3052	3067	3082	3097	3111	3126	3141	3156
25	3171	3186	3201	3216	3232	3247	3262	3277	3292	3307
26	3322	3337	3353	3368	3383	3398	3414	3429	3444	3460
27	3475	3490	3506	3521	3537	3552	3568	3583	3599	3614
28	3630	3645	3661	3676	3692	3708	3723	3739	3755	3770
29	3786	3802	3818	3833	3849	3865	3881	3897	3913	3928
30	3944	3960	3976	3992	4008	4024	4040	4056	4072	4088
31	4105	4121	4137	4153	4169	4185	4202	4218	4234	4250
32	4267	4283	4299	4316	4332	4348	4365	4381	4398	4414
33	4431	4447	4464	4480	4497	4513	4530	4546	4563	4580
34	4596	4613	4630	4646	4663	4680	4697	4713	4730	4747
35	4764	4781	4797	4814	4831	4848	4865	4882	4899	4916
36	4933	4950	4967	4984	5001	5018	5036	5053	5070	5087
37	5104	5121	5138	5155	5173	5190	5207	5225	5242	5259
38	5278	5295	5313	5330	5348	5365	5383	5400	5418	5436
39	5453	5471	5488	5505	5523	5541	5559	5576	5594	5612
40	5630	5647	5665	5683	5701	5719	5737	5755	5772	5790
41	5808	5826	5844	5862	5880	5898	5916	5935	5953	5971
42	5989	6007	6025	6043	6062	6080	6098	6116	6135	6153
43	6171	6190	6208	6226	6245	6263	6282	6300	6319	6337
44	6356	6374	6393	6411	6430	6448	6467	6486	6504	6523
45	6542	6560	6579	6598	6617	6635	6654	6673	6692	6711
46	6730	6749	6767	6786	6805	6824	6843	6862	6881	6900
47	6919	6938	6958	6977	6996	7015	7034	7053	7072	7092
48	7111	7130	7150	7169	7188	7208	7227	7246	7266	7285
49	7305	7324	7343	7363	7382	7402	7422	7441	7461	7480
50	7500	7520	7539	7559	7579	7598	7618	7638	7658	7677
51	7697	7717	7737	7757	7777	7796	7816	7836	7856	7876
52	7896	7916	7936	7956	7976	7997	8017	8037	8057	8077
53	8097	8117	8138	8158	8178	8198	8219	8239	8259	8280
54	8300	8320	8341	8361	8382	8402	8423	8443	8464	8484
55	8505	8525	8546	8566	8587	8608	8628	8649	8670	8690
56	8711	8732	8753	8773	8794	8815	8836	8857	8878	8899
57	8919	8940	8961	8982	9003	9024	9045	9066	9087	9109
58	9130	9151	9172	9193	9214	9235	9257	9278	9299	9320
59	9342	9363	9384	9406	9427	9448	9470	9491	9513	9534
60	9556	9577	9599	9620	9642	9663	9685	9706	9728	9750

For continuation to 170 feet, see table 15.

34

TABLE 13.—LEVEL CUTTINGS.

Roadway 18 feet wide, side-slopes ½ to 1.

Depth in ft.	·0	·1	·2	·3	·4	·5	·6	·7	·8	·9
	cu. yds.	cu. yds.	cu. yds.	cu. yds.	cu. yds.	cu. yds.	cu. yds.	cu. yds.	cu. yds.	cu. yds.
0		6·69	13·4	20·2	27·0	33·8	40·7	47·6	54·5	61·5
1	68·5	75·6	82·7	89·8	97·0	104·2	111·4	118·7	126·5	133·3
2	140·7	148·2	155·6	163·1	170·7	178·2	185·9	193·5	201·2	208·9
3	216·7	224·5	232·3	240·2	248·1	256·0	264·0	272·0	280·1	288·2
4	296·3	304·5	312·7	320·9	329·2	337·5	345·8	354·2	362·7	371·1
5	379·6	388·2	396·7	405·4	414·0	422·7	431·4	440·2	448·9	457·8
6	466·7	475·6	484·5	493·5	502·5	511·6	520·6	529·8	538·9	548·1
7	557·4	566·7	576·0	585·3	594·7	604·1	613·6	623·1	632·6	642·2
8	651·9	661·5	671·2	680·9	690·6	700·4	710·3	720·1	730·0	740·0
9	750·0	760·0	770·1	780·2	790·3	800·4	810·6	820·8	831·1	841·4
10	851·9	862·2	872·6	883·1	893·6	904·1	914·6	925·2	935·9	946·5
11	957·4	968·2	978·9	989·7	1000	1011	1022	1033	1044	1055
12	1067	1078	1089	1100	1111	1122	1134	1145	1156	1168
13	1180	1191	1203	1214	1226	1237	1249	1261	1272	1284
14	1296	1308	1320	1332	1344	1356	1368	1380	1392	1404
15	1417	1429	1442	1453	1466	1478	1490	1503	1515	1528
16	1541	1553	1566	1579	1591	1604	1617	1630	1642	1655
17	1668	1681	1694	1707	1721	1734	1747	1760	1773	1786
18	1800	1813	1827	1840	1853	1867	1880	1894	1908	1921
19	1935	1949	1963	1976	1990	2004	2018	2032	2046	2060
20	2074	2088	2102	2116	2131	2145	2159	2173	2188	2202
21	2217	2231	2246	2260	2275	2289	2304	2318	2333	2347
22	2363	2378	2393	2407	2422	2437	2452	2467	2482	2498
23	2513	2528	2543	2559	2574	2589	2605	2620	2635	2651
24	2667	2682	2698	2713	2729	2745	2761	2777	2793	2809
25	2825	2841	2857	2873	2889	2905	2921	2937	2954	2970
26	2985	3001	3019	3034	3051	3067	3084	3100	3117	3133
27	3150	3167	3183	3200	3217	3234	3251	3268	3284	3301
28	3318	3336	3353	3370	3387	3404	3421	3439	3456	3473
29	3491	3508	3526	3543	3561	3578	3596	3613	3631	3649
30	3667	3684	3702	3720	3738	3756	3774	3792	3810	3828
31	3846	3864	3882	3901	3919	3937	3956	3974	3992	4111
32	4030	4048	4067	4085	4104	4123	4141	4160	4179	4198
33	4217	4235	4254	4273	4292	4311	4330	4350	4369	4388
34	4407	4427	4446	4465	4485	4504	4524	4543	4563	4582
35	4602	4621	4641	4661	4680	4700	4720	4740	4760	4780
36	4800	4820	4840	4860	4880	4900	4921	4941	4961	4981
37	5002	5022	5043	5063	5084	5104	5125	5145	5166	5187
38	5207	5228	5249	5270	5291	5312	5332	5353	5374	5395
39	5417	5438	5459	5480	5501	5523	5544	5565	5587	5608
40	5630	5651	5673	5694	5716	5747	5759	5781	5803	5824
41	5846	5868	5890	5911	5934	5956	5979	6001	6023	6045
42	6067	6089	6111	6133	6156	6178	6201	6223	6246	6268
43	6291	6313	6336	6359	6381	6404	6427	6450	6472	6495
44	6519	6541	6564	6588	6611	6634	6657	6680	6703	6727
45	6750	6773	6797	6820	6844	6867	6891	6914	6938	6961
46	6985	7009	7033	7056	7080	7104	7128	7152	7176	7200
47	7224	7248	7272	7296	7321	7345	7369	7393	7418	7442
48	7467	7491	7515	7540	7565	7589	7614	7639	7663	7688
49	7713	7738	7763	7788	7812	7837	7862	7887	7913	7938
50	7963	7988	8013	8039	8064	8089	8115	8140	8166	8191
51	8217	8242	8268	8293	8319	8345	8371	8396	8422	8448
52	8474	8500	8526	8552	8578	8604	8630	8656	8683	8709
53	8735	8761	8788	8814	8841	8867	8894	8920	8947	8973
54	9000	9027	9053	9080	9107	9134	9161	9187	9214	9241
55	9269	9296	9323	9350	9377	9404	9431	9459	9486	9513
56	9541	9568	9596	9623	9651	9678	9706	9733	9761	9789
57	9817	9844	9872	9900	9928	9956	9984	10012	10040	10068
58	10096	10124	10153	10181	10209	10237	10266	10294	10323	10351
59	10380	10408	10437	10465	10494	10523	10551	10580	10609	10638
60	10667	10696	10724	10753	10782	10811	10841	10870	10899	10928

TABLE 14.—LEVEL CUTTINGS.

Roadway 28 *feet wide, side-slopes* ½ *to* 1.

Depth in feet	·0	·1	·2	·3	·4	·5	·6	·7	·8	·9
	cu. yds.	cu. yds.	cu. yds.	cu. yds.	cu. yds.	cu. yds.	cu. yds.	cu. yds.	cu. yds.	cu. yds.
0		10·4	20·8	31·3	41·8	52·3	62·9	73·5	84·1	94·8
1	105·6	126·3	137·1	147·9	158·8	169·7	180·7	191·6	202·7	213·7
2	214·8	225·9	237·1	248·3	259·5	270·8	282·1	293·5	304·9	316·3
3	327·8	339·3	350·8	362·4	374·0	385·6	397·3	409·0	420·8	432·6
4	444·4	456·3	468·2	480·2	492·1	504·2	516·2	528·3	540·4	552·6
5	564·8	577·1	589·3	601·6	614·0	626·4	638·8	651·3	663·8	676·3
6	688·9	701·5	714·1	726·8	739·5	752·3	765·1	777·9	790·8	803·7
7	816·7	829·7	842·7	858·8	868·9	882·1	895·3	908·5	921·8	935·1
8	948·1	961·5	974·9	988·3	1002	1015	1029	1042	1056	1070
9	1083	1097	1111	1125	1138	1152	1166	1180	1194	1208
10	1222	1236	1250	1265	1279	1293	1307	1322	1336	1350
11	1365	1379	1394	1408	1423	1437	1452	1467	1482	1496
12	1511	1526	1541	1556	1571	1586	1601	1616	1631	1646
13	1661	1676	1692	1707	1722	1737	1753	1768	1784	1799
14	1815	1830	1846	1862	1877	1893	1909	1925	1940	1956
15	1972	1988	2004	2020	2036	2052	2068	2085	2101	2117
16	2133	2150	2166	2182	2199	2215	2232	2248	2265	2281
17	2298	2315	2332	2348	2365	2382	2399	2416	2433	2450
18	2467	2484	2501	2518	2535	2552	2570	2587	2604	2621
19	2639	2656	2674	2691	2709	2726	2744	2762	2779	2797
20	2815	2833	2850	2868	2886	2904	2922	2940	2958	2976
21	2994	3013	3031	3049	3067	3086	3104	3122	3141	3159
22	3178	3196	3215	3233	3252	3271	3290	3308	3327	3346
23	3365	3384	3403	3422	3441	3460	3479	3498	3517	3536
24	3556	3575	3594	3613	3632	3652	3672	3692	3711	3731
25	3750	3770	3789	3809	3829	3849	3868	3888	3908	3928
26	3948	3968	3988	4008	4028	4049	4069	4089	4109	4130
27	4150	4170	4191	4211	4232	4252	4273	4294	4314	4335
28	4356	4376	4397	4418	4439	4460	4481	4502	4523	4544
29	4565	4586	4607	4628	4649	4671	4692	4713	4735	4756
30	4778	4799	4821	4843	4864	4886	4908	4929	4951	4973
31	4994	5016	5038	5060	5082	5104	5126	5148	5170	5192
32	5215	5237	5259	5281	5303	5326	5349	5371	5393	5416
33	5439	5461	5484	5507	5529	5552	5575	5598	5621	5644
34	5667	5690	5713	5736	5759	5782	5805	5828	5852	5875
35	5898	5922	5945	5968	5992	6015	6039	6063	6086	6110
36	6133	6157	6181	6205	6228	6252	6276	6300	6324	6348
37	6372	6396	6420	6445	6469	6493	6517	6542	6566	6590
38	6615	6639	6664	6688	6713	6738	6763	6787	6812	6837
39	6861	6886	6911	6936	6961	6986	7011	7036	7061	7086
40	7111	7136	7162	7187	7212	7237	7263	7288	7314	7339
41	7365	7390	7416	7442	7467	7493	7519	7545	7570	7596
42	7622	7648	7674	7700	7726	7752	7778	7805	7831	7857
43	7883	7910	7936	7962	7989	8015	8042	8068	8095	8121
44	8148	8175	8202	8228	8255	8282	8309	8336	8363	8390
45	8417	8444	8471	8498	8525	8552	8580	8607	8634	8661
46	8689	8716	8744	8771	8799	8826	8854	8882	8909	8937
47	8965	8993	9020	9048	9076	9104	9132	9160	9188	9216
48	9244	9273	9301	9329	9357	9386	9414	9442	9471	9499
49	9528	9556	9585	9613	9642	9671	9700	9728	9757	9786
50	9815	9844	9873	9902	9931	9960	9989	10018	10047	10076
51	10106	10135	10164	10193	10223	10252	10282	10311	10341	10370
52	10400	10430	10459	10489	10519	10549	10578	10608	10638	10668
53	10698	10728	10758	10788	10818	10849	10879	10909	10939	10970
54	11000	11030	11061	11091	11122	11152	11183	11213	11244	11275
55	11306	11336	11367	11398	11429	11460	11490	11521	11552	11583
56	11615	11646	11677	11708	11740	11771	11802	11833	11865	11896
57	11928	11959	11991	12022	12054	12086	12117	12149	12181	12213
58	12244	12276	12308	12340	12372	12404	12436	12468	12501	12533
59	12565	12597	12629	12662	12694	12726	12759	12791	12824	12856
60	12889	12922	12954	12987	13020	13052	13085	13118	13151	13184

36

TABLE 15.—LEVEL CUTTINGS.

Continuation of the foregoing Tables of Cubic Contents, to 170 feet of height or depth.

Height or Depth in feet.	TABLE 1	TABLE 2	TABLE 3	TABLE 4	TABLE 5	TABLE 6	TABLE 7	TABLE 8	TABLE 9	TABLE 10	TABLE 11	TABLE 12
	cu yd.	cu yd.	cu yd.	cu yd.	cu yd.	cu yd.	cu yd.	cu yd.	cu yd.	cu yd.	cu. yds.	cu. yds.
61	23835	26094	17848	21204	24739	31630	20107	23553	26998	33689	7512	9771
·5	24201	26479	18108	21610	25113	32117	20386	23888	27390	34394	7602	9880
62	24570	26867	18370	21930	25489	32607	20667	24226	27785	34904	7693	9989
·5	24942	27257	18634	22251	25868	33102	20949	24566	28183	35417	7784	10098
63	25317	27650	18900	22575	26250	33600	21233	24908	28583	35933	7875	10208
·5	25694	28046	19168	22901	26635	34102	21519	25253	28986	36454	7967	10319
64	26074	28444	19437	23230	27022	34607	21807	25600	29393	36978	8059	10430
·5	26457	28846	19708	23560	27413	35117	22097	25949	29801	37506	8152	10541
65	26843	29250	19981	23894	27806	35630	22389	26300	30213	38037	8245	10653
·5	27231	29657	20256	24229	28201	36146	22682	26654	30627	38572	8339	10765
66	27622	30067	20533	24567	23600	36667	22978	27011	31044	39111	8433	10878
·5	28016	30479	20812	24907	29001	37191	23275	27369	31464	39654	8528	10991
67	28413	30894	21093	25249	29406	37719	23574	27730	31887	40200	8623	11105
·5	28812	31313	21375	25594	29813	38250	23875	28093	32312	40750	8719	11219
68	29215	31733	21659	25941	30222	38785	24178	29459	32741	41304	8815	11333
·5	29620	32157	21945	26290	30635	39324	24482	28827	33172	41861	8911	11448
69	30028	32583	22233	26642	31050	39867	24789	29197	33605	42422	9008	11564
·5	30438	33013	22523	26996	31468	40413	25097	29569	34042	42987	9106	11680
70	30852	33444	22814	27352	31889	40963	25407	29944	34481	43556	9204	11796
·5	31268	33879	23108	27710	32313	41517	25719	30321	34924	44128	9302	11913
71	31687	34317	23404	28071	32739	42074	26033	30701	35369	44704	9401	12031
·5	32108	34757	23701	28434	33168	42635	26349	31083	35816	45283	9500	12148
72	32533	35200	24000	28800	33600	43200	26667	31467	36267	45867	9600	12267
·5	32960	35646	24301	29168	34035	43769	26986	31853	36720	46454	9700	12385
73	33390	36094	24604	29538	34472	44341	27307	32241	37176	47044	9801	12505
·5	33823	36546	24907	29910	34913	44917	27631	32632	37635	47639	9902	12624
74	34259	37000	25214	30285	35356	45496	27956	33026	38096	48237	10004	12744
·5	34697	37457	25522	30662	35801	46080	28282	33421	38561	48839	10106	12865
75	35139	37917	25833	31042	36250	46667	28611	33819	39028	49444	10208	12986
·5	35582	38379	26144	31423	36701	47257	28942	34219	39498	50054	10311	13108
76	36029	38844	26458	31807	37156	47852	29174	34622	39970	50667	10415	13230
·5	36479	39313	26774	32194	37613	48450	29608	35027	40446	51283	10519	13352
77	36931	39783	27092	32582	38072	49052	29944	35434	40924	51904	10623	13475
·5	37386	40257	27411	32973	38535	49657	30282	35843	41405	52528	10728	13599
78	37844	40733	27733	33367	39000	50267	30622	36256	41899	53156	10833	13722
·5	38305	41213	28056	33762	39468	50880	30964	36669	42375	53787	10939	13846
79	38768	41694	28381	34160	39939	51496	31307	37086	42865	54422	11045	13971
·5	39235	42179	28708	34560	40413	52117	31653	37504	43357	55061	11152	14096
80	39704	42667	29037	34963	40889	52741	32000	37926	43852	55704	11259	14222
81	40650	43650	29700	35775	41850	54000	32700	38775	44850	57000	11475	14475
82	41607	44644	30370	36596	42822	55274	33407	39633	45859	58311	11693	14730
83	42576	45650	31048	37427	43806	56563	34122	40501	46880	59637	11912	14986
84	43555	46667	31733	38267	44800	57867	34844	41378	47911	60978	12133	15244
85	44546	47694	32426	39116	45806	59185	35574	42264	48954	62333	12357	15505
86	45548	48733	33126	39974	46822	60519	36311	43159	50008	63704	12582	15767
87	46561	49783	33833	40842	47850	61867	37056	44064	51072	65089	12808	16031
88	47585	50844	34548	41718	48889	63230	37807	44978	52148	66489	13037	16296
89	48620	51917	35270	42605	49939	64607	38567	45901	53235	67904	13268	16564
90	49667	53000	36000	43500	51000	66000	39333	46833	54333	69333	13500	16833
91	50724	54094	36737	44405	52072	67407	40107	47775	55443	70778	13734	17105
92	51793	55200	37481	45319	53156	68830	40889	48726	56563	72237	13970	17378
93	52872	56317	38233	46242	54250	70267	41678	49686	57694	73711	14208	17653
94	53963	57444	38993	47174	55356	71719	42474	50655	58837	75200	14448	17930
95	55065	58583	39759	48116	56472	73185	43278	51634	59990	76704	14690	18208
96	56178	59733	40533	49067	57600	74667	44089	52622	61155	78222	14933	18489
97	57302	60894	41315	50027	58739	76163	44907	53619	62331	79756	15179	18771
98	58437	62067	42104	50996	59889	77674	45733	54626	63518	81304	15426	19056
99	59583	63250	42900	51975	61050	79200	46567	55642	64716	82867	15675	19342
100	60741	64444	43704	52963	62222	80741	47407	56667	65926	84444	15926	19630

TABLE 15.—LEVEL CUTTINGS—Continued.

Height or Depth in Feet.	Table 1	Table 2	Table 3	Table 4	Table 5	Table 6	Table 7	Table 8	Table 9	Table 10	Table 11	Table 12
101	61910	65649	44515	53960	63406	82297	48254	57700	67149	86036	16179	19919
102	63090	66868	45333	54967	64600	83806	49110	58744	68382	87643	16434	20211
103	64278	68096	46158	55983	65806	85452	49970	59797	69627	89258	16690	20505
104	65480	69335	47000	57008	67022	87052	50844	60859	70882	90903	16949	20800
105	66696	70586	47840	58042	68251	88606	51722	61930	72148	92555	17209	21097
106	67916	71846	48686	59085	69489	90296	52606	63011	73425	94222	17471	21396
107	69152	73118	49540	60138	70740	91941	53499	64100	74713	95904	17735	21697
108	70400	74400	50402	61200	72000	93600	54399	65199	76012	97600	18000	22000
109	71657	75695	51272	62272	73272	95275	55306	66308	77322	99311	18268	22305
110	72926	77000	52148	63352	74555	96963	56222	67426	78630	101037	18537	22611
111	74200	78319	53046	64442	75847	98666	57145	68552	79958	102777	18808	22919
112	75492	79640	53937	65541	77155	100385	58074	69687	81300	104533	19082	23229
113	76794	80988	54836	66650	78473	102118	59011	70832	82654	106303	19357	23541
114	78108	82336	55740	67767	79800	103866	59956	71986	84019	108089	19634	23855
115	79430	83696	56654	68894	81139	105610	60907	73149	85396	109888	19913	24171
116	80768	85068	57575	70030	82489	107408	61867	74322	86783	111704	20193	24488
117	82113	86451	58504	71176	83850	109200	62836	75504	88182	113532	20476	24807
118	83471	87845	59442	72330	85222	111007	63807	76695	89592	115378	20760	25128
119	84840	89250	60386	73494	86606	112830	64788	77900	91012	117237	21046	25452
120	86222	90666	61333	74660	88000	114667	65777	79111	92444	119111	21333	25777
121	87614	92097	62293	75849	89405	116519	66774	80330	93884	120999	21623	26103
122	89015	93536	63260	77041	90822	118386	67777	81558	95339	122901	21915	26432
123	90429	94985	64234	78242	92250	120267	68789	82795	96803	124818	22208	26763
124	91852	96446	65216	79452	93689	122164	69807	84044	98280	126750	22504	27095
125	93286	97919	66205	80671	95139	124075	70833	85300	99767	128703	22801	27430
126	94733	99402	67201	81900	96600	126002	71866	86568	101266	130666	23100	27766
127	96191	100896	68205	83138	98073	127943	72908	87840	102776	132643	23401	28104
128	97660	102401	69217	84385	99556	129898	73955	89122	104296	134636	23704	28444
129	99140	103917	70236	85642	101050	131869	75011	90418	105828	136643	24009	28786
130	100630	105444	71260	86908	102555	133854	76074	91722	107370	138666	24315	29129
131	102122	106984	72293	88183	104072	135855	77145	93034	108921	140702	24623	29475
132	103636	108535	73334	89467	105606	137869	78222	94356	110486	142754	24933	29822
133	105161	110096	74382	90761	107140	139898	79308	95686	112063	144821	25245	30171
134	106698	111668	75437	92063	108689	141942	80400	97026	113650	146903	25560	30522
135	108245	113251	76500	93375	110250	144002	81500	98375	115249	148999	25875	30875
136	109803	114846	77570	94697	111822	146075	82608	99732	116859	151111	26193	31230
137	111372	116451	78648	96027	113408	148164	83722	101100	118480	153237	26512	31586
138	112953	118068	79733	97367	115000	150267	84844	102477	120113	155378	26834	31944
139	114545	119695	80825	98716	116606	152386	85973	103863	121754	157534	27157	32304
140	116148	121333	81926	100078	118222	154519	87111	105259	123408	159704	27481	32666
141	117755	122986	83034	101443	119851	156646	88256	106664	125070	161888	27808	33030
142	119380	124645	84148	102820	121490	158829	89407	106079	126747	164089	28137	33395
143	121015	126317	85270	104206	123140	161027	90567	109501	128434	166303	28467	33763
144	122662	129000	86400	105601	124800	163200	91733	110034	130133	168533	28800	34132
145	124320	129694	87537	107005	126473	165406	92909	112375	131843	170777	29134	34504
146	125989	131400	88681	108419	128156	167629	94089	113820	133563	173037	29471	34877
147	127670	133117	89833	109842	129850	169866	95278	115286	135295	175311	29809	35252
148	129361	134845	90993	111274	131555	172118	96473	116755	137038	177600	30148	35629
149	131063	136582	92160	112715	133272	174385	97678	118234	138792	179903	30490	36008
150	132777	138333	93333	114106	135000	176666	98888	119722	140555	182222	30833	36389
151	134501	140003	94513	115626	136740	178962	100107	121220	142330	184555	31178	36771
152	136236	141866	95700	117094	138488	181274	101333	122726	144117	186904	31525	37155
153	137983	143650	96893	118574	140250	183600	102566	124241	145916	189266	31874	87541
154	139740	144444	98096	120062	142022	185941	103807	125766	147725	191644	32226	37929
155	141508	147249	99317	121560	143805	188296	105055	127300	149546	194036	32578	38319
156	143290	149067	100535	123066	145600	190666	106311	128844	151379	196444	32933	38711
157	145079	150894	101760	124582	147405	193052	107574	130397	153220	198866	33290	39104
158	146882	152733	102994	126107	149222	195452	108844	131959	155074	201303	33648	39500
159	148695	154583	104234	127641	151050	197867	110122	133529	156940	203755	34008	39897
160	150518	156444	105482	129185	152888	200296	111407	135108	158815	206222	34370	40296
161	152353	158316	106737	130738	154737	202740	112700	136700	160702	208703	34734	40697
162	154199	160200	108000	132300	156598	205200	114000	138300	162600	211200	35100	41100
163	156056	162094	109270	133873	158470	207674	115308	139908	164509	213710	35468	41505
164	157924	164000	110548	135454	160354	210163	116622	141526	166430	216237	35837	41911
165	159804	165916	111833	137046	162248	212667	117944	143153	168361	218777	36208	42319
166	161695	167844	113126	138645	164154	215186	119273	144788	170304	221334	36581	42730
167	163597	169783	114426	140255	166071	217720	120610	146434	172258	223906	36955	43142
168	165510	171732	115733	141873	167999	220268	121954	148089	174223	226489	37333	43556
169	167434	173693	117048	143502	169938	222831	123307	149753	176199	229089	37712	43972
170	169370	175666	118370	145130	171888	225409	124666	151424	178185	231704	38092	44389

TABLE 16.

Of cubic yards in a 100-foot station, to be added to, or subtracted from, the quantities in the preceding 15 tables, in case the excavations or embankments should be increased or diminished 2 feet in width. To be used only in rough estimates.

The most rapid method of performing this operation would be to add together the average heights, or depths, of the equivalent level cuttings or fillings of several consecutive 100-foot stations, and divide their sum by their number, for an average of them all. The number of cubic yards corresponding to this average height, or depth, when multiplied by the number of stations, will give the content of the entire length, *nearly*.

It would, however, be better for the assistant, in cases where the width of roadway differs from those in the preceding tables, to construct at once a new table; or else to use the diagrams and their tables in the manner described immediately following fig. 4½.

Cubic Yards in a length of 100 *feet; breadth* 2 *feet; and of different depths.*

Height or Depth in feet.	Cubic Yards.	Height or Depth in feet.	Cubic Yards.	Height or Depth in feet.	Cubic Yards.	Height or Depth in feet.	Cubic Yards.	Height or Depth in feet.	Cubic Yards.
·5	3·70	·5	152	·5	300	·5	448	·5	596
1	7·41	21	156	41	304	61	452	81	600
·5	11·1	·5	159	·5	307	·5	456	·5	604
2	14·6	22	163	42	311	62	459	82	607
·5	18·5	·5	167	·5	315	·5	463	·5	611
3	22·2	23	170	43	319	63	467	83	615
·5	25·9	·5	174	·5	322	·5	470	·5	619
4	29·6	24	178	44	326	64	474	84	622
·5	33·3	·5	181	·5	330	·5	478	·5	626
5	37·0	25	185	45	333	65	481	85	630
·5	40·7	·5	189	·5	337	·5	485	·5	633
6	44·4	26	193	46	341	66	489	86	637
·5	48·1	·5	196	·5	344	·5	493	·5	641
7	51·9		200	47	348	67	496	87	644
·5	55·6	·5	204	·5	352	·5	500	·5	648
8	59·3	28	207	48	356	68	504	88	652
·5	63·0	·5	211	·5	359	·5	507	·5	656
9	66·7	29	215	49	363	69	511	89	659
·5	70·4	·5	219	·5	367	.5	515	·5	663
10	74·1	30	222	50	370	70	519	90	667
·5	77·8	·5	226	·5	374	·5	522	·5	670
11	81·5	31	230	51	378	71	526	91	674
·5	85·2	·5	233	·5	381	·5	530	·5	678
12	88·9	32	237	52	385	72	533	92	681
·5	92·6	·5	241	·5	389	·5	537	·5	685
13	96·3	33	244	53	393	73	541	93	689
·5	100	·5	248	·5	396	·5	544	·5	693
14	104	34	252	54	400	74	548	94	696
·5	107	·5	256	·5	404	·5	552	·5	700
15	111	35	259	55	407	75	556	95	704
·5	115	·5	263	·5	411	·5	559	·5	707
16	119	36	267	56	415	76	563	96	711
·5	122	·5	270	·5	419	·5	567	·5	715
17	126	37	274	57	422	77	570	97	719
·5	130	·5	278	·5	426	·5	574	·5	722
18	133	38	281	58	430	78	578	98	726
·5	137	·5	285	·5	433	·5	581	·5	730
19	141	39	289	59	437	79	585	99	733
·5	144	·5	293	·5	441	·5	589	·5	737
20	148	40	296	60	444	80	593	100	741

TABLE 17.

This Table shows the number of cubic yards of Excavation, or Embankment, corresponding to *different areas of cross section*, and to a length of 100 feet. The areas are expressed in *square feet*.

This Table may be extended, by mentally changing the place of the decimal point; thus, the cubic yards corresponding to an area of 1100 square feet will be 10 times that of 110 square feet, or 407·4; those corresponding to an area of 8955 square feet will be 10 times that of 895·5 square feet, or 33167. If the number is not exactly divisible by 10, we may still take out the corresponding cubic yards with sufficient accuracy for practice, by using the nearest tabular number; thus, if we take 33167 for the cubic yards corresponding to an area of 8953 square feet, the error will be but about 8 cubic yards in 33000. A mean can, however, always be estimated in an instant by the eye, which will reduce the error to still less; thus it is seen at a glance, that 33157 is, in this instance, nearer than 33167.

Area. sq. ft.	Cubic Yards.	Area. sq. ft.	Cubic Yards.	Area. sq. ft.	Cubic Yards.	Area. sq. ft.	Cubic Yards.	Area. sq. ft.	Cubic Yards.	Area. sq. ft.	Cubic Yards.
1	3·70	45	106·7	89	329·6	·5	431·5	·5	513·0	·5	594·4
2	7·41	46	170·4	90	333·3	117	433·3	139	514·8	161	596·3
3	11·1	47	174·1	91	337·0	·5	435·2	·5	516·7	·5	598·2
4	14·8	48	177·8	92	340·7	118	437·0	140	518·5	162	600·0
5	18·5	49	181·5	93	344·4	·5	438·9	·5	520·4	·5	601·9
6	22·2	50	185·2	94	348·2	119	440·7	141	522·2	163	603·7
7	25·9	51	188·9	95	351·9	·5	442·6	·5	524·1	·5	605·6
8	29·6	52	192·6	96	355·6	120	444·4	142	525·9	164	607·4
9	33·3	53	196·3	97	359·3	·5	446·3	·5	527·8	·5	609·3
10	37·0	54	200·0	98	363·0	121	448·2	143	529·6	165	611·1
11	40·7	55	203·7	99	366·7	·5	450·0	·5	531·5	·5	613·0
12	44·4	56	207·4	100	370·4	122	451·9	144	533·3	166	614·8
13	48·1	57	211·1	·5	372·2	·5	453·7	·5	535·2	·5	616·7
14	51·9	58	214·8	101	374·1	123	455·6	145	537·0	167	618·5
15	55·6	59	218·5	·5	375·9	·5	457·4	·5	538·9	·5	620·4
16	59·3	60	222·2	102	377·8	124	459·3	146	540·7	168	622·2
17	63·0	61	225·9	·5	379·6	·5	461·1	·5	542·6	·5	624·1
18	66·7	62	229·6	103	381·5	125	463·0	147	544·4	169	625·9
19	70·4	63	233·3	·5	383·3	·5	464·8	·5	546·3	·5	627·8
20	74·1	64	237·0	104	385·2	126	466·7	148	548·2	170	629·6
21	77·8	65	240·7	·5	387·0	·5	468·5	·5	550·0	·5	631·5
22	81·5	66	244·4	105	388·9	127	470·4	149	551·9	171	633·3
23	85·2	67	248·2	·5	390·7	·5	472·2	·5	553·7	·5	635·2
24	88·9	68	251·9	106	392·6	128	474·1	150	555·6	172	637·0
25	92·6	69	255·6	·5	394·4	·5	475·9	·5	557·4	·5	638·9
26	96·3	70	259·3	107	396·3	129	477·8	151	559·3	173	640·7
27	100·0	71	263·0	·5	398·2	·5	479·6	·5	561·1	·5	642·6
28	103·7	72	266·7	108	400·0	130	481·5	152	563·0	174	644·4
29	107·4	73	270·4	·5	401·9	·5	483·3	·5	564·8	·5	646·3
30	111·1	74	274·1	109	403·7	131	485·2	153	566·7	175	648·2
31	114·8	75	277·8	·5	405·6	·5	487·0	·5	568·5	·5	650·0
32	118·5	76	281·5	110	407·4	132	488·9	154	570·4	176	651·9
33	122·2	77	285·2	·5	409·3	·5	490·7	·5	572·2	·5	653·7
34	125·9	78	288·9	111	411·1	133	492·6	155	574·1	177	655·6
35	129·6	79	292·6	·5	413·0	·5	494·4	·5	575·9	·5	657·4
36	133·3	80	296·3	112	414·8	134	496·3	156	577·8	178	659·3
37	137·0	81	300·0	·5	416·7	·5	498·2	·5	579·6	·6	661·1
38	140·7	82	303·7	113	418·5	135	500·0	157	581·5	179	663·0
39	144·4	83	307·4	·5	420·4	·5	501·9	·5	583·3	·5	664·8
40	148·2	84	311·1	114	422·2	136	503·7	158	585·2	180	666·7
41	151·9	85	314·8	·5	424·1	·5	505·6	·5	587·0	·5	668·5
42	155·6	86	318·5	115	425·9	137	507·4	159	588·9	181	670·4
43	159·3	87	322·2	·5	427·8	·5	509·3	·5	590·7	·5	672·2
44	163·0	88	325·9	116	429·6	138	511·1	160	592·6	182	674·1

TABLE 17.

Area. sq. ft.	Cubic Yards.	Area. sq. ft.	Cubic Yards.	Area. sq. ft.	Cubic Yards.	Area. sq. ft.	Cubic Yards.	Area. sq. ft.	Cubic Yards.	Area. sq. ft.	Cubic Yards.
·5	675·9	·5	783·3	·5	890·7	·5	998·2	·5	1105·6	·5	1213·0
183	677·8	212	785·2	241	892·6	270	1000·0	299	1107·4	328	1214·8
·5	679·6	·5	787·0	·5	894·4	·5	1001·9	·5	1109·3	·5	1216·7
184	681·5	213	788·9	242	896·3	271	1003·7	300	1111·1	329	1218·5
·5	683·3	·5	790·7	·5	898·2	·5	1005·6	·5	1113·0	·5	1220·4
185	685·2	214	792·6	243	900·0	272	1007·4	301	1114·8	330	1222·2
·5	687·0	·5	794·4	·5	901·9	·5	1009·3	·5	1116·7	·5	1224·1
186	688·9	215	796·3	244	903·7	273	1011·1	302	1118·5	331	1225·9
·5	690·7	·5	798·2	·5	905·6	·5	1013·0	·5	1120·4	·5	1227·8
187	692·6	216	800·0	245	907·4	274	1014·8	303	1122·2	332	1229·6
·5	694·4	·5	801·9	·5	909·3	·5	1016·7	·5	1124·1	·5	1231·5
188	696·3	217	803·7	246	911·1	275	1018·5	304	1125·9	333	1233·3
·5	698·2	·5	805·6	·5	913·0	·5	1020·4	·5	1127·8	·5	1235·2
189	700·0	218	807·4	247	914·8	276	1022·2	305	1129·6	334	1237·0
·5	701·9	·5	809·3	·5	916·7	·5	1024·1	·5	1131·5	·5	1238·9
190	703·7	219	811·1	248	918·5	277	1025·9	306	1133·3	335	1240·7
·5	705·6	·5	813·0	·5	920·4	·5	1027·8	·5	1135·2	·5	1242·6
191	707·4	220	814·8	249	922·2	278	1029·6	307	1137·0	336	1244·4
·5	709·3	·5	816·7	·5	924·1	·5	1031·5	·5	1138·9	·5	1246·3
192	711·1	221	818·5	250	925·9	279	1033·3	308	1140·7	337	1248·2
·5	713·0	·5	820·4	·5	927·8	·5	1035·2	·5	1142·6	·5	1250·0
193	714·8	222	822·2	251	929·6	280	1037·0	309	1144·4	338	1251·9
·5	716·7	·5	824·1	·5	931·5	·5	1038·9	·5	1146·3	·5	1253·7
194	718·5	223	825·9	252	933·3	281	1040·7	310	1148·2	339	1255·6
·5	720·4	·5	827·8	·5	935·2	·5	1042·6	·5	1150·0	·5	1257·4
195	722·2	224	829·6	253	937·0	282	1044·4	311	1151·9	340	1259·3
·5	724·1	·5	831·5	·5	938·9	·5	1046·3	·5	1153·7	·5	1261·1
196	725·9	225	833·3	254	940·7	283	1048·2	312	1155·6	341	1263·0
·5	727·8	·5	835·2	·5	942·6	·5	1050·0	·5	1157·4	·5	1264·8
197	729·6	226	837·0	255	944·4	284	1051·9	313	1159·3	342	1266·7
·5	731·5	·5	838·9	·5	946·3	·5	1053·7	·5	1161·1	·5	1268·5
198	733·3	227	840·7	256	948·2	285	1055·6	314	1163·0	343	1270·4
·5	735·2	·5	842·6	·5	950·0	·5	1057·4	·5	1164·8	·5	1272·2
199	737·0	228	844·4	257	951·9	286	1059·3	315	1166·7	344	1274·1
·5	738·9	·5	846·3	·5	953·7	·5	1061·1	·5	1168·5	·5	1275·9
200	740·7	229	848·2	258	955·6	287	1063·0	316	1170·4	345	1277·8
·5	742·6	·5	850·0	·5	957·4	·5	1064·8	·5	1172·2	·5	1279·6
201	744·4	230	851·9	259	959·3	288	1066·7	317	1174·1	346	1281·5
·5	746·3	·5	853·7	·5	961·1	·5	1068·5	·5	1175·9	·5	1283·3
202	748·2	231	855·6	260	963·0	289	1070·4	318	1177·8	347	1285·2
·5	750·0	·5	857·4	·5	964·8	·5	1072·2	·5	1179·6	·5	1287·0
203	751·9	232	859·3	261	966·7	290	1074·1	319	1181·5	348	1288·9
·5	753·7	·5	861·1	·5	968·5	·5	1075·9	·5	1183·3	·5	1290·7
204	755·6	233	863·0	262	970·4	291	1077·8	320	1185·2	349	1292·6
·5	757·4	·5	864·8	·5	972·2	·5	1079·6	·5	1187·0	·5	1294·4
205	759·3	234	866·7	263	974·1	292	1081·5	321	1188·9	350	1296·3
·5	761·1	·5	868·5	·5	975·9	·5	1083·3	·5	1190·7	·5	1298·2
206	763·0	235	870·4	264	977·8	293	1085·2	322	1192·6	351	1300·0
·5	764·8	·5	872·2	·5	979·6	·5	1087·0	·3	1194·4	·5	1301·9
207	766·7	236	874·1	265	981·5	294	1088·9	323	1196·3	352	1303·7
·5	768·5	·5	875·9	·5	983·3	·5	1090·7	·5	1198·2	·5	1305·6
208	770·4	237	877·8	266	985·2	295	1092·6	324	1200·0	353	1307·4
·5	772·2	·5	879·6	·5	987·0	·5	1094·4	·5	1201·9	·5	1309·3
209	774·1	238	881·5	267	988·9	296	1096·3	325	1203·7	354	1311·1
·5	775·9	·5	883·3	·5	990·7	·5	1098·2	·5	1205·6	·5	1313·0
210	777·8	239	885·2	268	992·6	297	1100·0	326	1207·4	355	1314·8
·5	779·6	·5	887·0	·5	994·4	·5	1101·9	·5	1209·3	·5	1316·7
211	781·5	240	888·9	269	996·3	298	1103·7	327	1211·1	356	1318·6

Table 17.

Area sq. ft.	Cubic Yards.	Area sq. ft.	Cubic Yards.	Area sq. ft.	Cubic Yards.	Area sq. ft.	Cubic Yards.	Area sq. ft.	Cubic Yards.	Area sq. ft.	Cubic Yard.	Area sq. ft.	Cubic Yards.
·5	1320·4	·5	1427·8	·5	1535·2	·5	1642·6	·5	1750·0	·5	1857·4
357	1322·2	386	1429·6	415	1537·0	444	1644·4	473	1751·9	502	1859·3
·5	1324·1	·5	1431·5	·5	1538·9	·5	1646·3	·5	1753·7	·5	1861·1
358	1325·9	387	1433·3	416	1540·7	445	1648·2	474	1755·6	503	1863·0
·5	1327·8	·5	1435·2	·5	1542·6	·5	1650·0	·5	1757·4	·5	1864·8
359	1329·6	388	1437·0	417	1544·4	446	1651·9	475	1759·3	504	1866·7
·5	1331·5	·5	1439·9	·5	1546·3	·5	1653·7	·5	1761·1	·5	1868·5
360	1333·3	389	1440·7	418	1548·2	447	1655·6	476	1763·0	505	1870·4
·5	1335·2	·5	1442·6	·5	1550·0	·5	1657·4	·5	1764·8	·5	1872·2
361	1337·0	390	1444·4	419	1551·9	448	1659·3	477	1766·7	506	1874·1
·5	1338·9	·5	1446·3	·5	1553·7	·5	1661·1	·5	1768·5	·5	1875·9
362	1340·7	391	1448·2	420	1555·6	449	1663·0	478	1770·4	507	1877·8
·5	1342·6	·5	1450·0	·5	1557·4	·5	1664·8	·5	1772·2	·5	1879·6
363	1344·4	392	1451·9	421	1559·3	450	1666·7	479	1774·1	508	1881·5
·5	1346·3	·5	1453·7	·5	1561·1	·5	1668·5	·5	1775·9	·5	1883·3
364	1348·2	393	1455·6	422	1563·0	451	1670·4	480	1777·8	509	1885·2
·5	1350·0	·5	1457·4	·5	1564·8	·5	1672·2	·5	1779·6	·5	1887·0
365	1351·9	394	1459·3	423	1566·7	452	1674·1	481	1781·5	510	1888·9
·5	1353·7	·5	1461·1	·5	1568·5	·5	1675·9	·5	1783·3	·5	1890·7
366	1355·6	395	1463·0	424	1570·4	453	1677·8	482	1785·2	511	1892·6
·5	1357·4	·5	1464·8	·5	1572·2	·5	1679·6	·5	1787·0	·5	1894·4
367	1359·3	396	1466·7	425	1574·1	454	1681·5	483	1788·9	512	1896·3
·5	1361·1	·5	1468·5	·5	1575·9	·5	1683·3	·5	1790·7	·5	1898·2
368	1363·0	397	1470·4	426	1577·8	455	1685·2	484	1792·6	513	1900·0
·5	1364·8	·5	1472·2	·5	1579·6	·5	1687·0	·5	1794·4	·5	1901·9
369	1366·7	398	1474·1	427	1581·5	456	1688·9	485	1796·3	514	1903·7
·5	1368·5	·5	1475·9	·5	1583·3	·5	1690·7	·5	1798·2	·5	1905·6
370	1370·4	399	1477·8	428	1585·2	457	1692·6	486	1800·0	515	1907·4
·5	1372·2	·5	1479·6	·5	1587·0	·5	1694·4	·5	1801·9	·5	1909·3
371	1374·1	400	1481·5	429	1588·9	458	1696·3	487	1803·7	516	1911·1
·5	1375·9	·5	1483·3	·5	1590·7	·5	1698·2	·5	1805·6	·5	1913·0
372	1377·8	401	1485·2	430	1592·6	459	1700·0	488	1807·4	517	1914·8
·5	1379·6	·5	1487·0	·5	1594·4	·5	1701·9	·5	1809·3	·5	1916·7
373	1381·5	402	1488·9	431	1596·3	460	1703·7	489	1811·1	518	1918·5
·5	1383·3	·5	1490·7	·5	1598·2	·5	1705·6	·5	1813·0	·5	1920·4
374	1385·2	403	1492·6	432	1600·0	461	1707·4	490	1814·8	519	1922·2
·5	1387·0	·5	1494·4	·5	1601·9	·5	1709·3	·5	1816·7	·5	1924·1
375	1388·9	404	1496·3	433	1603·7	462	1711·1	491	1818·5	520	1925·9
·5	1390·7	·5	1498·2	·5	1605·6	·5	1713·0	·5	1820·4	·5	1927·8
376	1392·6	405	1500·0	434	1607·4	463	1714·8	492	1822·2	521	1929·6
·5	1394·4	·5	1501·9	·5	1609·3	·5	1716·7	·5	1824·1	·5	1931·5
377	1396·3	406	1503·7	435	1611·1	464	1718·5	493	1825·9	522	1933·3
·5	1398·2	·5	1505·6	·5	1613·0	·5	1720·4	·5	1827·8	·5	1935·2
378	1400·0	407	1507·4	436	1614·8	465	1722·2	494	1829·6	523	1937·0
·5	1401·9	·5	1509·3	·5	1616·7	·5	1724·1	·5	1831·5	·5	1938·9
379	1403·7	408	1511·1	437	1618·5	466	1725·9	495	1833·3	524	1940·7
·5	1405·6	·5	1513·0	·5	1620·4	·5	1727·8	·5	1835·2	·5	1942·6
380	1407·4	409	1514·8	438	1622·2	467	1729·6	496	1837·0	525	1944·4
·5	1409·3	·5	1516·7	·5	1624·1	·5	1731·5	·5	1838·9	·5	1946·3
381	1411·1	410	1518·5	439	1625·9	468	1733·3	497	1840·7	526	1948·2
·5	1413·0	·5	1520·4	·5	1627·8	·5	1735·2	·5	1842·6	·5	1950·0
382	1414·8	411	1522·2	440	1629·6	469	1737·0	498	1844·4	527	1951·9
·5	1416·7	·5	1524·1	·5	1631·5	·5	1738·9	·5	1846·3	·5	1953·7
383	1418·5	412	1525·9	441	1633·3	470	1740·7	499	1848·2	528	1955·6
·5	1420·4	·5	1527·8	·5	1635·2	·5	1742·6	·5	1850·0	·5	1957·4
384	1422·2	413	1529·6	442	1637·0	471	1744·4	500	1851·9	529	1959·3
·5	1424·1	·5	1531·5	·5	1638·9	·5	1746·3	·5	1853·7	·5	1961·1
385	1425·9	414	1533·3	443	1640·7	472	1748·2	501	1855·6	530	1963·0

TABLE 17.

Area. sq. ft.	Cubic Yards.	Area. sq. ft.	Cubic Yards.	Area. sq. ft.	Cubic Yards.	Area. sq. ft.	Cubic Yards.	Area. sq. ft.	Cubic Yards.	Area. sq. ft.	Cubic Yards.
·5	1964·8	·5	2072·2	·5	2179·6	·5	2287·0	·5	2394·4	·5	2501·9
531	1966·7	500	2074·1	589	2181·5	618	2288·9	647	2396·3	676	2503.7
·5	1968·5	·5	2075·9	·5	2183·3	·5	2290·7	·5	2398·2	·5	2505·6
532	1970·4	561	2077·8	590	2185·2	619	2292·6	648	2400·0	677	2507·4
·5	1972·2	·5	2079·6	·5	2187·0	·5	2294·4	·5	2401·9	·5	2509·3
533	1974·1	562	2081·5	591	2188·9	620	2296·3	649	2403·7	678	2511·1
·5	1975·9	·5	2083·3	·5	2190·7	·5	2298·2	·5	2405·6	·5	2513·0
534	1977·8	563	2085·2	592	2192·6	621	2300·0	650	2407·4	679	2514·8
·5	1979·6	·5	2087·0	·5	2194·4	·5	2301·9	·5	2409·3	·5	2516·7
535	1981·5	564	2088·9	593	2196·3	622	2303·7	651	2411·1	680	2518·5
·5	1983·3	·5	2090·7	·5	2198·2	·5	2305·6	·5	2413·0	·5	2520·4
536	1985·2	565	2092·6	594	2200·0	623	2307·4	652	2414·8	681	2522·2
·5	1987·0	·5	2094·4	·5	2201·9	·5	2309·3	·5	2416·7	·5	2524·1
537	1988·9	566	2096·3	595	2203·7	624	2311·1	653	2418·5	682	2525·9
·5	1990·7	·5	2098·2	·5	2205·6	·5	2313·0	·5	2420·4	·5	2527·8
538	1992·6	567	2100·0	596	2207·4	625	2314·8	654	2422·2	683	2529·6
·5	1994·4	·5	2101·9	·5	2209·3	·5	2316·7	·5	2424·1	·5	2531·5
539	1996·3	568	2103·7	597	2211·1	626	2318·5	655	2425·9	684	2533·3
·5	1998·2	·5	2105·6	·5	2213·0	·5	2320·4	·5	2427.8	·5	2535·2
540	2000·0	569	2107·4	598	2214·8	627	2322·2	656	2429·6	685	2537·0
·5	2001·9	·5	2109·3	·5	2216·7	·5	2324·1	·5	2431·5	·5	2538·9
541	2003·7	570	2111·1	599	2218·5	628	2325·9	657	2433·3	686	2540·7
·5	2005·6	·5	2113·0	·5	2220·4	·5	2327·8	·5	2435·2	·5	2542·6
542	2007·4	571	2114·8	600	2222·2	629	2329·6	658	2437·0	687	2544·4
·5	2009·3	·5	2116·7	·5	2224·1	·5	2331·5	·5	2438·9	·5	2546·3
543	2011·1	572	2118·5	601	2225·9	630	2333·3	659	2440·7	688	2548·2
·5	2013·0	·5	2120·4	·5	2227·8	·5	2335·2	·5	2442·6	·5	2550·0
544	2014·8	573	2122·2	602	2229·6	631	2337·0	660	2444·4	689	2551·9
·5	2016·7	·5	2124·1	·5	2231·5	·5	2338·9	·5	2446·3	·5	2553·7
545	2018·5	574	2125·9	603	2233·3	632	2340·7	661	2448·2	690	2555·6
·5	2020·4	·5	2127·8	·5	2235·2	·5	2342·6	·5	2450·0	·5	2557·4
546	2022·2	575	2129·6	604	2237·0	633	2344·4	662	2451·9	691	2559·3
·5	2024.1	·5	2131·5	·5	2238·9	·5	2346·3	·5	2453·7	·5	2561·1
547	2025·9	576	2133·3	605	2240·7	634	2348·2	663	2455·6	692	2563·0
·5	2027·8	·5	2135·2	·5	2242·6	·5	2350·0	·5	2457·4	·5	2564·8
548	2029·6	577	2137·0	606	2244·4	635	2351·9	664	2459·3	693	2566·7
·5	2031·5	·5	2138·9	·5	2246·3	·5	2353·7	·5	2461·1	·5	2568·5
549	2033·3	578	2140·7	607	2248·2	636	2355·6	665	2463·0	694	2570·4
·5	2035·2	·5	2142·6	·5	2250·0	·5	2357·4	·5	2464·8	·5	2572·2
550	2037·0	579	2144·4	608	2251·9	637	2359·3	666	2466·7	695	2574·1
·5	2038·9	·5	2146·3	·5	2253·7	·5	2361·1	·5	2468·5	·5	2575·9
551	2040·7	580	2148·2	609	2255·6	638	2363·0	667	2470·4	696	2577·8
·5	2042·6	·5	2150·0	·5	2257·4	·5	2364·8	·5	2472·2	·5	2579·6
552	2044·4	581	2151·9	610	2259·3	639	2366·7	668	2474·1	697	2581·5
·5	2046·3	·5	2153·7	·5	2261·1	·5	2368·5	·5	2475·9	·5	2583·3
553	2048·2	582	2155·6	611	2263·0	640	2370·4	669	2477·8	698	2585·2
·5	2050·0	·5	2157·4	·5	2264·8	·5	2372·2	·5	2479·6	·5	2587·0
554	2051·9	583	2159·3	612	2266·7	641	2374·1	670	2481·5	699	2588·9
·5	2053·7	·5	2161·1	·5	2268·5	·5	2375·9	·5	2483·3	·5	2590·7
555	2055·6	584	2163·0	613	2270·4	642	2377·8	671	2485·2	700	2592·6
·5	2057·4	·5	2164·8	·5	2272·2	·5	2379·6	·5	2487·0	·5	2594·4
556	2059·3	585	2166·7	614	2274·1	643	2381·2	672	2488·9	701	2596·3
·5	2061·1	·5	2168·5	·5	2275·9	·5	2383·3	·5	2490·7	·5	2598·2
557	2063·0	586	2170·4	615	2277·8	644	2385·2	673	2492·6	702	2600·0
·5	2064·8	·5	2172·2	·5	2279·6	·5	2387·0	·5	2494·4	·5	2601·9
558	2066·7	587	2174·1	616	2281·5	645	2388·9	674	2496·3	703	2603·7
·5	2068·5	·5	2175·9	·5	2283·3	·5	2390·7	·5	2498·2	·5	2605·6
559	2070·4	588	2177·8	617	2285·2	646	2392·6	675	2500·0	704	2607·4

43
TABLE 17.

Area. sq. ft.	Cubic Yards.	Area. sq. ft.	Cubic Yards.	Area. sq. ft.	Cubic Yards.	Area. sq. ft.	Cubic Yards.	Area. sq. ft.	Cubic Yards.	Area. sq. ft.	Cubic Yards.
·5	2609·3	·5	2716·7	·5	2824·1	·5	2931·5	·5	3038·9	·5	3146·3
705	2611·1	734	2718·5	763	2825·9	792	2933·3	821	3040·7	850	3148·2
·5	2613·0	·5	2720·4	·5	2827·8	·5	2935·2	·5	3042·6	·5	3150·0
706	2614·8	735	2722·2	764	2829·6	793	2937·0	822	3044·4	851	3151·9
·5	2616·7	·5	2724·1	·5	2831·5	·5	2939·9	·5	3046·3	·5	3153·7
707	2618·5	736	2725·9	765	2833·3	794	2940·7	823	3048·2	852	3155·6
·5	2620·4	·5	2727·8	·5	2835·2	·5	2942·6	·5	3050·0	·5	3157·4
708	2622·2	737	2729·6	766	2837·0	795	2944·4	824	3051·9	853	3159·3
·5	2624·1	·5	2731·5	·5	2838·9	·5	2946·3	·5	3053·7	·5	3161·1
709	2625·9	738	2733·3	767	2840·7	796	2948·2	825	3055·6	854	3163·0
·5	2627·8	·5	2735·2	·5	2842·6	·5	2950·0	·5	3057·4	·5	3164·8
710	2629·6	739	2737·0	768	2844·4	797	2951·9	826	3059·3	855	3166·7
·5	2631·5	·5	2738·9	·5	2846·3	·5	2953·7	·5	3061·1	·5	3168·5
711	2633·3	740	2740·7	769	2848·2	798	2955·6	827	3063·0	856	3170·4
·5	2635·2	·5	2742·6	·5	2850·0	·5	2957·4	·5	3064·8	·5	3172·2
712	2637·0	741	2744·4	770	2851·9	799	2959·3	828	3066·7	857	3174·1
·5	2638·9	·5	2746·3	·5	2853·7	·5	2961·1	·5	3068·5	·5	3175·9
713	2640·7	742	2748·2	771	2855·6	800	2963·0	829	3070·4	858	3177·8
·5	2642·6	·5	2750·0	·5	2857·4	·5	2964·8	·5	3072·2	·5	3179·6
714	2644·4	743	2751·9	772	2859·3	801	2966·7	830	3074·1	859	3181·5
·5	2646·3	·5	2753·7	·5	2861·1	·5	2968·5	·5	3075·9	·5	3183·3
715	2648·2	744	2755·6	773	2863·0	802	2970·4	831	3077·8	860	3185·2
·5	2650·0	·5	2757·4	·5	2864·8	·5	2972·2	·5	3079·6	·5	3187·0
716	2651·9	745	2759·3	774	2866·7	803	2974·1	832	3081·5	861	3188·9
·5	2653·7	·5	2761·1	·5	2868·5	·5	2975·9	·5	3083·3	·5	3190·7
717	2655·6	746	2763·0	775	2870·4	804	2977·8	833	3085·2	862	3192·6
·5	2657·4	·5	2764·8	·5	2872·2	·5	2979·6	·5	3087·0	·5	3194·4
718	2659·3	747	2766·7	776	2874·1	805	2981·5	834	3088·9	863	3196·3
·5	2661·1	·5	2768·5	·5	2875·9	·5	2983·3	·5	3090·7	·5	3198·2
719	2663·0	749	2770·4	777	2877·8	806	2985·2	835	3092·6	864	3200·0
·5	2664·8	·5	2772·2	·5	2879·6	·5	2987·0	·5	3094·4	·5	3201·9
720	2666·7	749	2774·1	778	2881·5	807	2988·9	836	3096·3	865	3203·7
·5	2668·5	·5	2775·9	·5	2883·3	·5	2990·7	·5	3098·2	·5	3205·6
721	2670·4	750	2777·8	779	2885·2	808	2992·6	837	3100·0	866	3207·4
·5	2672·2	·5	2779·6	·5	2887·0	·5	2994·4	·5	3101·9	·5	3209·3
722	2674·1	751	2781·5	780	2888·9	809	2996·3	838	3103·7	867	3211·1
·5	2675·9	·5	2783·3	·5	2890·7	·5	2998·2	·5	3105·6	·5	3213·0
723	2677·8	752	2785·2	781	2892·6	810	3000·0	839	3107·4	868	3214·8
·5	2679·6	·5	2787·0	·5	2894·4	·5	3001·9	·5	3109·3	·5	3216·7
724	2681·5	753	2788·9	782	2896·3	811	3003·7	840	3111·1	869	3218·5
·5	2683·3	·5	2790·7	·5	2898·2	·5	3005·6	·5	3113·0	·5	3220·4
725	2685·2	754	2792·6	783	2900·0	812	3007·4	841	3114·8	870	3222·2
·5	2687·0	·5	2794·4	·5	2901·9	·5	3009·3	·5	3116·7	·5	3224·1
726	2688·9	755	2796·3	784	2903·7	813	3011·1	842	3118·5	871	3225·9
·5	2690·7	·5	2798·2	·5	2905·6	·5	3013·0	·5	3120·4	·5	3227·8
727	2692·6	756	2800·0	785	2907·4	814	3014·8	843	3122·2	872	3229·6
·5	2694·4	·5	2801·9	·5	2909·3	·5	3016·7	·5	3124·1	·5	3231·5
728	2696·3	757	2803·7	786	2911·1	815	3018·5	844	3125·9	873	3233·3
·5	2698·2	·5	2805·6	·5	2913·0	·5	3020·4	·5	3127·8	·5	3235·2
729	2700·0	758	2807·4	787	2914·8	816	3022·2	845	3129·6	874	3237·0
·5	2701·9	·5	2809·3	·5	2916·7	·5	3024·1	·5	3131·5	·5	3239·9
730	2703·7	759	2811·1	788	2918·5	817	3025·9	846	3133·3	875	3240·7
·5	2705·6	·5	2813·0	·5	2920·4	·5	3027·8	·5	3135·2	·5	3242·6
731	2707·4	760	2814·8	789	2922·2	818	3029·6	847	3137·0	876	3244·4
·5	2709·3	·5	2816·7	·5	2924·1	·5	3031·5	·5	3138·9	·5	3245·3
732	2711·1	761	2818·5	790	2925·9	819	3033·3	848	3140·7	877	3249·2
·5	2713·0	·5	2820·4	·5	2927·8	·5	3035·2	·5	3142·6	·5	3250·0
733	2714·8	762	2822·2	791	2929·6	820	3037·0	849	3144·4	878	3251·9

44
Table 17.

Area. sq. ft.	Cubic Yards.	Area. sq. ft.	Cubic Yards.	Area. sq. ft.	Cubic Yards.	Area. sq. ft.	Cubic Yards.	Area. sq. ft.	Cubic Yards.	Area. sq. ft.	Cubic Yards.
·5	3253·7	899	3329·6	·5	3405·6	940	3481·5	·5	3557·4	981	3633·3
879	3255·6	·5	3331·5	920	3407·4	·5	3483·3	961	3559·3	·5	3635·2
·5	3257·4	900	3333·3	·5	3409·3	941	3495·2	·5	3561·1	982	3637·0
880	3259·3	·5	3335·2	921	3411·1	·5	3487·0	962	3563·0	·5	3638·9
·5	3261·1	901	3337·0	·5	3413·0	942	3488·9	·5	3564·8	983	3640·7
881	3263·0	·5	3338·9	922	3414·8	·5	3490·7	963	3566·7	·5	3642·6
·5	3264·8	902	3340·7	·5	3416·7	943	3492·6	·5	3568·5	984	3644·4
882	3266·7	·5	3342·6	923	3418·5	·5	3494·4	964	3570·4	·5	3646·3
·5	3268·5	903	3344·4	·5	3420·4	944	3496·3	·5	3572·2	985	3648·2
883	3270·4	·5	3346·3	924	3422·2	·5	3499·2	965	3574·1	·5	3650·0
·5	3272·2	904	3348·2	·5	3424·1	945	3500·0	·5	3575·9	986	3651·9
884	3274·1	·5	3350·0	925	3425·9	·5	3501·9	966	3577·8	·5	3653·7
·5	3275·9	905	3351·9	·5	3427·8	946	3503·7	·5	3579·6	987	3655·6
985	3277·8	·5	3353·7	926	3429·6	·5	3505·6	967	3581·5	·5	3657·4
·5	3279·6	906	3355·6	·5	3431·5	947	3507·4	·5	3583·3	988	3659·3
886	3281·5	·5	3357·4	927	3433·3	·5	3509·3	968	3585·2	·5	3661·1
·5	3283·3	907	3359·3	·5	3435·2	948	3511·1	·5	3587·0	989	3663·0
887	3285·2	·5	3361·1	928	3437·0	·5	3513·0	969	3588·9	·5	3664·8
·5	3287·0	908	3363·0	·5	3438·9	949	3514·8	·5	3590·7	990	3666·7
888	3288·9	·5	3364·8	929	3440·7	·5	3516·7	970	3592·6	·5	3668·5
·5	3290·7	909	3366·7	·5	3442·6	950	3519·5	·5	3594·4	991	3670·4
889	3292·6	·5	3368·5	930	3444·4	·5	3520·4	971	3596·3	·5	3672·2
·5	3294·4	910	3370·4	·5	3446·3	951	3522·2	·5	3598·2	992	3674·1
890	3296·3	·5	3372·2	931	3448·2	·5	3524·1	972	3600·0	·5	3675·9
·5	3298·2	911	3374·1	·5	3450·0	952	3525·9	·5	3601·9	993	3677·8
891	3300·0	·5	3375·9	932	3451·9	·5	3527·8	973	3603·7	·5	3679·6
·5	3301·9	912	3377·8	·5	3453·7	953	3529·6	·5	3605·6	994	3681·5
892	3303·7	·5	3379·6	933	3455·6	·5	3531·5	974	3607·4	·5	3683·3
·5	3305·6	913	3381·5	·5	3457·4	954	3533·3	·5	3609·3	995	3685·2
893	3307·4	·5	3383·3	934	3459·3	·5	3535·2	975	3611·1	·5	3687·0
·5	3309·3	914	3385·2	·5	3461·1	955	3537·0	·5	3613·0	996	3688·9
894	3311·1	·5	3387·0	935	3463·0	·5	3538·9	976	3614·8	·5	3690·7
·5	3313·0	915	3388·9	·5	3464·8	956	3540·7	·5	3616·7	997	3692·6
895	3314·8	·5	3390·7	936	3466·7	·5	3542·6	977	3618·5	·5	3694·4
·5	3316·7	916	3392·6	·5	3468·5	957	3544·4	·5	3620·4	998	3696·3
896	3318·5	·5	3394·4	937	3470·4	·5	3546·3	978	3622·2	·5	3698·2
·5	3320·4	917	3396·3	·5	3472·2	958	3548·2	·5	3624·1	999	3700·0
897	3322·2	·5	3398·2	938	3474·1	·5	3550·0	979	3625·9	·5	3701·9
·5	3324·1	918	3400·0	·5	3475·9	959	3551·9	·5	3627·8	1000	3703·7
898	3325·9	·5	3401·9	939	3477·8	·5	3553·7	980	3629·6		
·5	3327·8	919	3403·7	·5	3479·6	960	3555·6	·5	3631·5		

To PREPARE A TABLE, T (below), OF LEVEL CUTTINGS, FOR EVERY $\frac{1}{10}$ OF A FOOT OF HEIGHT OR DEPTH.

Let the fig. represent the cutting; or, if inverted, the filling; in which the horizontal lines are supposed to be $\frac{1}{10}$ foot apart.
First calculate the area in square feet, of the layer $a\,b\,c\,o$, adjoining the roadway $a\,b$. Then find how many cubic yards that area gives in a distance of 100 feet. These cubic yards we will call Y; they form the first amount to be put into the table T.

Next calculate the area in square feet of the triangle $a\,n\,o$. Multiply this area by 4. Find how many cubic yards this increased area gives in a distance of 100 feet. Or they will be found ready calculated a little farther on. We will call them y. This is all the preparation that is needed before commencing the table.

Example. Let the roadbed $a\,b$ be 18 feet, and the side-slopes $1\frac{1}{2}$ to 1, as in our preceding table and diagram No. V. Then for the area of $a\,b\,c\,o$: since the side-slopes are $1\frac{1}{2}$ to 1; and $s\,t$ is ·1 foot; $c\,o$ must be 18·3 feet; and the mean length of $a\,b\,c\,o$ must be 18·15 feet. Consequently the area is $18·15 \times ·1 = 1·815$ square feet; which, in a distance of 100 feet, gives 181·5 cubic feet; which is equal to
$$\frac{181·5}{27} = 6·7222 \text{ cubic yards; or Y.}$$

Next, as to the triangle $a\,n\,o$: its height $a\,n$ being ·1 foot, and its base $n\,o$ ·15 feet; its area $= \dfrac{·1 \times ·15}{2} = \dfrac{·015}{2} = ·0075$ square ft. This multiplied by 4, gives ·03 square feet; which, in a distance of 100 feet, gives $·03 \times 100 = 3$ cubic feet; which is equal to $\dfrac{3}{27} = ·1111$ cubic yard; or y.

Having thus found Y and y, proceed to make out the table in the manner following, which is so plain as to require no explanation. The work should be tested about every 5 feet, by calculating the area of the full depth arrived at; multiply it by 100, and divide the product by 27 for the cubic yards. The cubic yards thus found should agree with the table.

Y......6·7222...	......Y.	6·7222	·1
y...... ·1111			
6·8333		6·8333	
y;...... ·1111		13·5555	·2
6·9444		6·9444	
y...... ·1111		20·5000	·3
7·0555		7·0555	.
y...... ·1111		27·5555	·4
7·1666		7·1666	
y...... ·1111		34·7222	·5
. 7·2777		7·2777	
		42·0000	·6

TABLE T.

Height. Feet.	Cub. Yds.
·1.......	6·72 Y.
·2.......	13·6
·3.......	20·5
·4.......	27·6
·5.......	34·7
·6.......	42·0
&c	

The following table contains y, ready calculated for different side-slopes. It plainly remains the same for all widths of roadbed.

Side-slope.	y.	Side-slope.	y.
¼ to 1	·0185	1¾ to 1	·1296
½ to 1	·0370	2 to 1	·1482
¾ to 1	·0556	2¼ to 1	·1667
1 to 1	·0741	2½ to 1	·1852
1¼ to 1	·0926	3 to 1	·2222
1½ to 1	·1111	4 to 1	·2963

Mr. John R. Hudson, C. E., author of "Tables for Calculating the Cubic Contents of Excavations and Embankments,"* suggests the following modification of the foregoing method. We insert it with his permission. Its advantage is that it enables us to perform all, or nearly all, of the additions *mentally*.

Prepare a sheet of paper to contain the table, ruling it into eleven columns as in fig. A below, and as in our tables 1 to 14, and giving it as many horizontal lines as there are feet in the greatest depth for which the table is to be used. Leave enough space between the lines to allow of writing pencil figures between the contents which finally constitute the table.†

Find the contents for depths from ·1 foot to ·9 foot by our foregoing method, and write them in ink in their proper places in the table, as in the first line of fig. A, for which we have taken, as an example, a roadway 18 feet wide, with side-slopes of 1½ to 1, as in the preceding example.

Calling the content (64·5 cubic yards in the above example) for a depth of ·9 foot, "M", add to it mentally the quantity "P" (1 cubic yard in this case) corresponding to the given side-slope (1½ to 1) in table X below. Write their sum, M + P, (65·5) in pencil under the content (6·7) for a depth of ·1 foot. To M + P (65·5) add mentally another "P" (1 cubic yard) and write their sum M + 2 P (66·5) in pencil under the content (13·6) for a depth of .2 foot, *i. e.* in the next space to the right. Again, adding P (1 cubic yard) to this last sum, we obtain M + 3 P (67·5) which we write in pencil under the content (20·5) for a depth of ·3 foot; and so on until we write M + 9 P (73·5) under the content M (64·5) for a depth of ·9 foot.

Now call this 9 P (9 cubic yards) "S", and add it in turn to each of the sums (M + P, M + 2 P, etc.) just written in pencil in the first line, writing each new sum in pencil in the lower part of the *second* line and always *one column to the left* of the quantity

---

* John Wiley & Sons, Publishers, New York, second edition, 1886.

† In figs. A and B the heavy figures represent ink, and the light ones pencil.

(M + P, M + 2 P, etc.) from which it was obtained. Thus, taking M + P (65·5) which we had pencilled in the space corresponding to a depth of ·1 foot, and adding S (9 cubic yards) to it, we write their sum M + S + P (74·5) in pencil in the space *next below it and to the left*, namely, in the space for the content for a depth of one foot. Thus each sum, so obtained, is written in pencil in the space corresponding to a depth ·9 foot greater than that of the space which contains the quantity (M + P, etc.) from which it was obtained. The last sum thus obtained from the first line is M + S + 9 P = M + 2 S (82.5), which we pencil in the space corresponding to a depth of 1·8 feet. We next add S (9 cubic yards) to M + S + P (74·5) which we had pencilled in the space for a depth of one foot, and pencil their sum (83·5) in the space for a depth of 1·9 feet (·9 foot greater, as before) which we find *at the other end of the second line*. Proceed now with the second line, as with the first, writing the sums in pencil in the third line and adding S continuously. Then do the same with the following lines. When this is finished, the table will present the appearance indicated by fig. A.

We have selected for illustration a side-slope of 1½ to 1, because with that slope the quantity P, used in obtaining the contents written in ink in the first line, is 1 cubic yard, and this renders the explanation simpler. It will be seen at once that each pencilled quantity is greater by "P" than that for a depth ·1 foot less, and hence that *with said side-slope* of 1½ to 1, it would be easier to discard the use of "S", which is = 9 P, and simply make each pencilled quantity throughout the table 1 cubic yard greater than the preceding one. But for most other slopes P contains a fraction, while S remains a whole number; and in such cases it will therefore be better to use S and proceed as directed above.

Having thus obtained the pencilled figures for the entire table, as shown in fig. A (which, in order to save space, we have extended only to a depth of 2 feet) proceed to find the final quantities to be entered in the table, and write them in ink in their places as in fig. B.

To do this, add together the content (6·7) for depth ·1 foot, and the pencilled sum M + P (65·5) immediately below it, obtaining as their sum the content (72·2) for a depth of 1 foot (·9 foot greater) and write it in ink in its proper place in the table as shown in heavy figures in fig. B. Proceeding thus with the other quantities in the first line, we add each content (written in ink) to the sum immediately below it (written in pencil) and enter the new sum in ink in its proper place in the next line below and in the next column to the left, *i. e.* in the space corresponding to a depth ·9 foot greater. The last content thus obtained from the first line will be that (138 cubic yards) for a depth of 1·8 feet. For a depth of 1·9 feet we now add the content (72·2) for a depth of 1 foot, to the sum, M + P + S (74·5) pencilled immediately below it, and obtain the volume (146·7) for a depth of 1·9 feet and write it in ink in its proper place at the other end of the second line. Continue this process until the table is completed, as in fig. B.

TABLE X.

Slope.	$P\left(=\frac{8}{9}\right).$	$S(=9\ P).$
¼ to 1	.16667	1.5
½ "	.33333	3
1 "	.66667	6
1½ "	1.	9
2 "	1.33333	12
2½ "	1.66667	15
3 "	2.	18
4 "	2.66667	24

FIG. A.
Roadway 18 feet wide, side-slopes 1½ to 1.

Depth in feet.	.0	.1	.2	.3	.4
0	Cubic yards.	Cubic yards. 6.7 M+P 65.5	Cubic yards. 13.6 M+2 P 66.5	Cubic yards. 20.5 M+3 P 67.5	Cubic yards. 27.6 M+4 P 68.5
1	M+S+P 74.5	M+S+2 P 75.5	M+S+3 P 76.5	M+S+4 P 77.5	M+S+5 P 78.5
2	M+2 S+2 P 84.5	M+2 S+3 P 85.5	M+2 S+4 P 86.5	M+2 S+5 P 87.5	M+2 S+6 P 88.5

Depth in feet.	.5	.6	.7	.8	.9
0	Cubic yards. 34.7 M+5 P 69.5	Cubic yards. 42.0 M+6 P 70.5	Cubic yards. 49.4 M+7 P 71.5	Cubic yards. 56.9 M+8 P 72.5	Cubic yards. M 64.5 M+9 P=M+S 73.5
1	M+S+6 P 79.5	M+S+7 P 80.5	M+S+8 P 81.5	M+2 S 82.5	M+2 S+P 83.5
2	M+2 S+7 P 89.5	M+2 S+8 P 90.5	M+3 S 91.5	M+3 S+P 92.5	M+3 S+2 P 93.5

FIG. B.
Roadway 18 feet wide, side-slopes 1½ to 1.

Depth in feet.	.0	.1	.2	.3	.4	.5	.6	.7	.8	.9
	Cubic yards.	Cubic yards.	Cubic yards.	Cubic yards.	Cubic yards.	Cubic yards.	Cubic yards.	Cubic yards.	Cubic yards.	Cubic yards.
0		6.7	13.6	20.5	27.6	34.7	42.0	49.4	56.9	64.5
		65.5	66.5	67.5	68.5	69.5	70.5	71.5	72.5	73.5
1	72.2	80.1	88.0	96.1	104.2	112.5	120.9	129.4	138.0	146.7
	74.5	75.5	76.5	77.5	78.5	79.5	80.5	81.5	82.5	83.5
2	155.6	164.5	173.6	182.7	192.0	201.4	210.9	220.5	230.2	240.1
	84.5	85.5	86.5	87.5	88.5	89.5	90.5	91.5	92.5	93.5

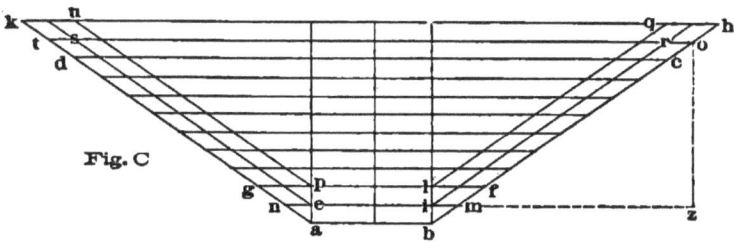

Fig. C

P is the sum of the contents, in cubic yards, of two prisms, $i\,m\,o\,r$ and $e\,n\,t\,s$, fig. C, each of which is 100 feet long and has for its cross section a parallelogram, as $i\,m\,o\,r$, whose depth $o\,z$ is ·9 foot, and whose width $i\,m$ is ·1 foot × the ratio of the horizontal to the vertical measure of the slope. Thus, if the slope is 2 to 1, P is =

$$\frac{2 \times \cdot 1 \times 2 \times \cdot 9 \times 100}{27} = \frac{2 \times 2 \times 9}{27} = \tfrac{2}{3} \times 2 = \tfrac{4}{3} = 1\cdot 333,\text{ as in}$$

table X.

S is = 9 P.

Or, for any slope, calling the vertical measure of the slope 1, as usual:

P = ⅔ × horizontal measure of the slope.
S = 6 ×   "   "   "   "

The difference, $n\,m\,o\,t$, between the section $a\,b\,m\,n$ ·1 foot deep and the section $a\,b\,o\,t$ 1 foot deep, is made up of $e\,i\,r\,s$ (= $a\,b\,c\,d$ or a section ·9 foot deep) and the two parallelograms $i\,m\,o\,r$ and $e\,n\,t\,s$; or content for depth 1 foot

    = content for depth ·1 foot
    +   "   "   " ·9 foot (= M)
    + twice content for parallelogram $i\,m\,o\,r$
    = content for depth ·1 foot + M + P.

Similarly, the difference $gfhk$, between section $abfg$, ·2 foot deep and $abhk$ 1·1 feet deep, is made up of $plqu$ (= $abcd$ ·9 foot deep) and the two parallelograms $lfhq$ and $pgku$, together = 2 P; or,

    content for depth 1·1 feet
        = content for depth ·2 foot + M + 2 P

and so on; or,
    content for any depth X
        = content for depth Y, ·9 foot less than X,
        + M
        + P × number of tenths in depth Y.

TO CALCULATE BEFOREHAND THE CUBIC CONTENTS OF BORROW-PITS.

The method of doing this most readily, is based upon the following rule, for finding the contents of any frustum,* as A, B, or C, of a *square*† prism, no matter how the two ends may be inclined with regard to each other; or whether one, or neither of them, is parallel to the original base of the prism.

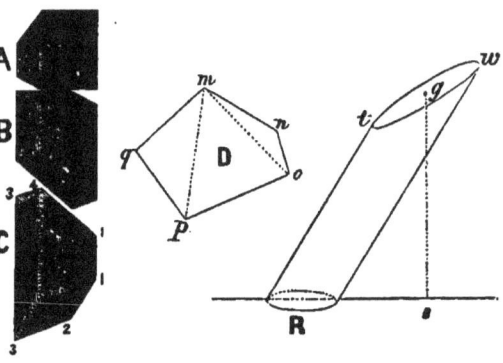

* Generally misspelt "frustrum."

† This rule applies also to frustums of prisms whose cross-sections perpendicular to the parallel sides are either any triangle, any parallelogram, or any *regular* polygon of any number of sides whatever. The parallel edges must in every case be added together, and divided by *their number*, for a mean length. In a square prism this number is 4; in a triangular one 3, &c.

If the frustum is that of an *irregular* 4-sided, or polygonal prism, it must be considered as made up of a number of triangular prisms, which must be calculated separately and added together for the total content. Thus, let D represent the cross-section of an irregular prism, perpendicular to its parallel sides. From any angle, as m, draw lines mo, mp, dividing it into triangles. Calculate the area of each triangle separately, and find the solidity of each triangular frustum, by first measuring the lengths of its three parallel edges, and then obtaining their mean length, which multiply by their triangular area.

THE SOLIDITY OR CONTENT OF ANY FRUSTUM WHATEVER, OF ANY PRISM, WHETHER REGULAR OR IRREGULAR, OR OF A CYLINDER, may be found thus: Consider either of its two ends, as R, to be its base, and find the area of that end. Find also the centre of gravity, g, of the other end, and measure the perpendicular height gs, of the frustum. Multiply the area of the base just found, by this height; the product will be the content.

The centre of gravity, g, may be found by cutting out a figure of the end tw from pasteboard, and balancing it in two directions over a sharp edge.

RULE. *Measure and add together the four parallel edges,* (as 11, 22, 33, 44, *of the frustum* C.) *Divide their sum by* 4, *for a mean length. Multiply this mean length by the area of the frustum at right angles to said parallel edges. The product will be the required content.*

To apply this rule to borrow-pits, the surface of the ground is first staked out in squares. If the surface is sloping or irregular, the tape-line must be held horizontally, while laying out the sides of the squares. When the ground is very irregular, these squares should be of such a size that each one of them may without material error be considered to be a plane surface, either horizontal or sloping. The depth of the horizontal bottom of the pit, below a certain given mark, or datum, being first determined on, levels are then taken at all the corners of the squares to ascertain the depth of digging at each corner. These depths plainly give the lengths of the four parallel edges of each frustum; each frustum may then be calculated separately, and the whole added together.

COST OF EARTHWORK.

ARTICLE 1.

It is advisable to pay for this kind of work by the cubic yard of *excavation* only, instead of allowing separate prices for excavation and embankment. By this means we get rid of the difficulty of measurements, as well as the controversies and lawsuits which often attend the determination of the allowance to be made for the settlement or subsidence of the embankments.

It is, moreover, our opinion that justice to the contractor should lead to the English practice of paying the laborers by the cubic yard, instead of by the day. Experience fully proves that when laborers are scarce and wages high, men can scarcely be depended upon to do three-fourths of the work which they readily accomplish when wages are low, and when fresh hands are waiting to be hired in case any are discharged. The contractor is thus placed at the mercy of his men. The writer has known the most satisfactory results to attend a system of task-work, accompanied by liberal premiums for all overwork. By this means the interests of the laborers are identified with that of the contractor, and every man takes care that the others shall do their fair share of the task.

Ellwood Morris, C. E., of Philadelphia, was, we believe, the first person who properly investigated the elements of cost of earthwork, and reduced them to such a form as to enable us to calculate the total with a considerable degree of accuracy. He published his results in the Journal of the Franklin Institute in 1841. His paper forms the basis on which, with some variations, we shall consider the matter, and on which we shall extend it to wheelbarrows, as well as to carts. Throughout this paper we speak of a cubic yard considered only as solid in its place, or before it is loosened for removal. It is scarcely necessary to add that the various items can of course only be regarded as tolerably close approximations, or averages. As before stated, the men do less work when wages are high, and more when they are low. A great deal besides depends on the skill, observation, and energy of the contractor and his superintendents. It is no unusual thing to see two contractors working at the same prices, in precisely similar material, where one is making money, and the other losing it, from a want of tact in the proper distribution of his forces, keeping his roads in order, having his carts and barrows well filled, &c., &c. Uncommonly long spells of wet weather may seriously affect the cost of executing earthwork, by making it more difficult to loosen, load, or empty; besides keeping the roads in bad order for hauling.

The aggregate cost of excavating and removing earth is made up by the following items, namely:

1st. *Loosening the earth ready for the shovellers.*
2d. *Loading it by shovels into the carts or barrows.*
3d. *Hauling, or wheeling it away, including emptying and returning.*
4th. *Spreading it out into successive layers on the embankment.*
5th. *Keeping the hauling-road for carts, or the plank gangways for barrows, in good order.*
6th. *Wear, sharpening, depreciation, and interest on cost of tools.*
7th. *Superintendence and water-carriers.*
8th. *Profit to the contractor.*

We will consider these items a little in detail, basing our calculations on the assumption that common labor costs $1 per day of 10 working hours. The results in our tables must therefore be increased or diminished in about the same proportion as common labor costs more or less than this.

ARTICLE 2.

LOOSENING THE EARTH READY FOR THE SHOVELLERS.—This is generally done either by ploughs or by picks; more cheaply by the first. A plough with two horses, and two men to manage them, at $1 per day for labor, 75 cents per day for each horse, and 37 cents per day for plough, including harness, wear, repairs, &c., or a total of $3.87, will loosen, of strong heavy soils, from 200 to 300 cubic yards a day, at from 1·93 to 1·29 cents per yard; or of ordinary loam, from 400 to 600 cubic yards a day, at from ·97 to ·64 of a cent per yard. Therefore, as an ordinary average, we may assume the actual cost to the contractor for loosening by the plough, as follows: strong heavy soils, 1·5 cents; common loam, ·8 cent; light sandy soils, ·4 cent. Very stiff pure clay, or obstinate cemented gravel, may be set down at 2·5 cents; they require three or four horses.

By the pick, a fair day's work is about 14 yards of stiff pure clay, or of cemented gravel; 25 yards of strong heavy soils; 40 yards of common loam; 60 yards of light sandy soils—all measured in place; which, at $1 per day for labor, gives, for stiff clay, 7 cents; heavy soils, 4 cents; loam, 2·5 cents; light sandy soil, 1·666 cents. Pure sand requires but very little labor for loosening; ·5 of a cent will cover it.

ARTICLE 3.

SHOVELLING THE LOOSENED EARTH INTO CARTS.—The amount shovelled per day depends partly upon the weight of the material, but more upon so proportioning the number of pickers and of carts to that of shovellers, as not to keep the latter waiting for either material or carts. In fairly regulated gangs, the shovellers into carts are not actually engaged in shovelling for more than six-tenths of their time, thus being unoccupied but four-tenths of it; while, under bad management, they lose considerably more than one-half of it. A shoveller can readily load into a cart one-third of a cubic yard measured in place (and which is an average working cart-load), of sandy soil, in five minutes; of loam, in six minutes; and of any of the heavy soils, in seven minutes. This would give, for a day of 10 working hours, 120 loads, or 40 cubic yards of light sandy soil; 100 loads, or 33⅓ cubic yards of loam; or 86 loads, or 28·7 yards of the heavy soils. But from these amounts we must deduct four-tenths for time necessarily lost; thus reducing the actual working quantities to 24 yards of light sandy soil, 20 yards of loam, 17·2 yards of the heavy soils. When the shovellers do less than this, there is some mismanagement.

Assuming these as fair quantities, then, at $1 per day for labor, the actual cost to the contractor for shovelling per cubic yard measured in place, will be, for sandy soils, 4·167 cents; loam, 5 cents; heavy soils, clays, &c., 5·81 cents.

In practice, the carts are not usually loaded to any less extent with the heavier soils than with the lighter ones. Nor, indeed, is there any necessity for so doing, inasmuch as the difference of weight of a cart and one-third of a cubic yard of the various soils is too slight to need any attention; especially when the cart-road is kept in good order, as it will be by any contractor who understands his own interest. Neither is it necessary to modify the load on account of any *slight inclinations* which may occur in the grading of roads. An earth-cart weighs by itself about ½ a ton.

ARTICLE 4.

HAULING AWAY THE EARTH, DUMPING OR EMPTYING, AND RETURNING TO RELOAD.—The average speed of horses in hauling is about 2⅓ miles per hour, or 200 feet per minute; which is equal to 100 feet of trip each way; or to 100 feet of *lead*, as the distance to which *the earth is hauled* is technically called. Beside this, there is a loss of about four minutes in every trip, whether long or short, in waiting to load, dumping, turning, &c. Hence, every trip will occupy as many minutes as there are lengths of 100 feet each in the lead; and four minutes beside. Therefore, to find the number of trips per day over any given average lead, we divide the number of minutes in a working day by the sum of 4 added to the number of 100-feet lengths contained in the distance to which the earth has to be removed; that is,

$$\frac{\text{The number (600) of minutes in a working day}}{4 + \text{the number of 100-feet lengths in the lead}} = \text{the number of trips, or loads removed per day, per cart.}$$

And since $\frac{1}{3}$ of a cubic yard measured before being loosened, makes an average cart-load, the number of loads, divided by 3, will give the number of cubic yards removed per day by each cart; and the cubic yards divided into the total expense of a cart per day, will give the cost per cubic yard for hauling.

Remark.—When removing loose *rock*, which requires more time for loading, say,

$$\frac{\text{No. of minutes (600) in a working day.}}{6 + \text{No. of 100-feet lengths of lead.}} = \frac{\text{No. of loads removed}}{\text{per day, per cart.}}$$

In leads of ordinary length one driver can attend to 4 carts; which, at $1 per day, is 25 cents per cart. When labor is at $1 per day the expense of a horse is usually about 75 cents; and that of the cart, including harness, tar, repairs, &c., 25 cents, making the total daily cost per cart $1.25. The expense of the horse is the same on Sundays and on rainy days, as when at work; and this consideration is included in the 75 cents. Some contractors employ a greater number of drivers, who also help to load the carts, so that the expense is about the same in either case.

Example. How many cubic yards of loam, measured in the cut, can be hauled by a horse and cart in a day of 10 working-hours, (600 minutes,) the lead, or length of haul of earth being 1000 feet, (or 10 lengths of 100 feet,) and what will be the expense to the contractor for hauling, per cubic yard, assuming the total cost of cart, horse, and driver, at $1·25?

Here, $\dfrac{600 \text{ minutes}}{4 + 10 \text{ lengths of 100 feet,}} = \dfrac{600}{14} = 43 \text{ loads.}$ And $\dfrac{43 \text{ loads}}{3} =$ 14·3 cubic yards. And $\dfrac{125 \text{ cents}}{14·3 \text{ cub. yds.}} = 8·74$ cents per cubic yard.

In this manner the 2d and 3d columns of the following tables have been calculated.

ARTICLE 5.

SPREADING, OR LEVELLING OFF THE EARTH INTO REGULAR THIN LAYERS ON THE EMBANKMENT.—A bankman will spread from 50 to 100 cub. yds. per day of either common loam, or any of the heavier soils, clays, &c., depending on their dryness. This, at $1 per day, is 1 to 2 cents per cubic yard; and we may assume $1\frac{1}{2}$ cents as a fair average for such soils; while 1 cent will suffice for light sandy soils.

This expense for spreading is saved when the earth is either dumped over the end of the embankment, or is wasted; still, about $\frac{1}{4}$ cent per yard should be allowed in either case for keeping the dumping-places clear and in order.

ARTICLE 6.

KEEPING THE CART-ROAD IN GOOD ORDER FOR HAULING.—No ruts or puddles should be allowed to remain unfilled; rain should at once be led off by shallow ditches; and the road be carefully kept in good order; otherwise the labor of the horses, and the wear of carts, will be *very greatly* increased. It is usual to allow so much per cubic yard for road repairs; but we suggest so much per cubic yard, per 100 feet of lead; say $\frac{1}{10}$ of a cent.

ARTICLE 7.

WEAR, SHARPENING, AND DEPRECIATION OF PICKS AND SHOVELS.—Experience shows that about $\frac{1}{4}$ of a cent per cubic yard will cover this item.

SUPERINTENDENCE AND WATER-CARRIERS.—These expenses will vary with local circumstances; but we agree with Mr. Morris, that $1\frac{1}{2}$ cents per cubic yard will, under ordinary circumstances, cover both of them. An allowance of about $\frac{1}{4}$ cent may in justice be added for extra trouble in digging the side-ditches; levelling off the bottom of the cut to grade; and general trimming up. In very light cuttings this may be increased to $\frac{1}{2}$ cent per cubic yard.

At $\frac{1}{4}$ cent, all the items in this Article amount to 2 cents per cubic yard of cut.

ARTICLE 8.

PROFIT TO THE CONTRACTOR.—This may generally be set down at from 6 to 15 per cent., according to the magnitude of the work, the risks incurred, and various incidental circumstances. Out of this item the contractor generally has to pay clerks, storekeepers, and other agents, as well as the expenses of shanties, &c.; although these are in most cases repaid by the profits of the stores; and by the rates of boarding and lodging paid to the contractors by the laborers.

ARTICLE 9.

A knowledge of the foregoing items enables us to calculate with tolerable accuracy the cost of removing earth. For example, let it be required to ascertain the cost per cubic yard of excavating common loam, measured in place; and of removing it into embankment, with an average haul or lead of 1000 feet; the wages of laborers being $1 per day of 10 working hours; a horse 75 cents a day; and a cart 25 cents. One driver to four carts.

		Cents.
Here we have cost of loosening, say by pick, Art. 2, per cubic yard, say,		2·50
Loading into carts, Art. 3, " "		5·00
Hauling 1000 feet, as calculated previously in example, Art. 4, "		8·74
Spreading into layers, Art. 5, "		1·50
Keeping cart-road in repair, Art. 6, 10 lengths of 100 ft.,		1·00
Various items in Art. 7, . . .		. 2·00
Total cost to contractor, ...		20·74
Add contractor's profit, say 10 per cent.,	.	2·074
Total cost per cubic yard to the company, .	.	22·814

It is easy to construct a table like the following, of costs per cubic yard, for different lengths of lead. Columns 2 and 3 are first obtained by the Rule in Article 4; then to each amount in column 3 is added the *variable* quantity of $\frac{1}{10}$ of a cent for every 100 feet length of lead, for keeping the road in order; and the *constant* quantity (for any given kind of soil) composed of the prices per cubic yard, for loosening, loading, spreading, or wasting, &c., either taken from the preceding Articles; or modified to suit particular circumstances. In this manner the tables have been prepared.

BY CARTS.—LABOR $1 PER DAY, OF 10 WORKING HOURS.

Length of Load, or distance to which the earth is hauled, in feet.	Number of cubic yards in place, hauled per day by each cart.	Cost per cubic yard in place, for hauling and emptying only.	COMMON LOAM. TOTAL COST PER CUBIC YARD, EXCLUSIVE OF PROFIT TO CONTRACTOR.				STRONG HEAVY SOILS. TOTAL COST PER CUBIC YARD, EXCLUSIVE OF PROFIT TO CONTRACTOR.			
			Picked and Spread.	Picked and Wasted.	Ploughed and Spread.	Ploughed and Wasted.	Picked and Spread.	Picked and Wasted.	Ploughed and Spread.	Ploughed and Wasted.
Feet.	cub. yds.	cts.	cts.	cts.	cts.	cts.	cts.	cts.	cts.	cts.
25	47·0	2·66	13·69	12·44	11·99	10·74	16·00	14·75	13·50	12·25
50	44·4	2·81	13·86	12·61	12·16	10·91	16·17	14·92	13·67	12·42
75	42·1	2·97	14·05	12·80	12·35	11·10	16·36	15·11	13·86	12·61
100	40·0	3·12	14·22	12·97	12·52	11·27	16·53	15·28	14·03	12·78
150	36·4	3·43	14·58	13·33	12·88	11·63	16·89	15·64	14·39	13·14
200	33·3	3·75	14·95	13·70	13·25	12·00	17·26	16·01	14·76	13·51
300	28·6	4·37	15·67	14·42	13·97	12·72	17·98	16·73	15·48	14·23
400	25·0	5·00	16·40	15·15	14·70	13·45	18·71	17·46	16·21	14·96
500	22·2	5·63	17·13	15·88	15·43	14·18	19·44	18·19	16·94	15·69
600	20·0	6·25	17·85	16·60	16·15	14·90	20·16	18·91	17·66	16·41
700	18·2	6·87	18·57	17·32	16·87	15·62	20·88	19·63	18·38	17·13
800	16·7	7·48	19·28	18·03	17·58	16·33	21·59	20·34	19·09	17·84
900	15·4	8·12	19·92	18·67	18·22	16·97	22·23	20·98	19·73	18·48
1000	14·3	8·74	20·74	19·49	19·04	17·79	23·05	21·80	20·55	19·30
1100	13·3	9·40	21·50	20·25	19·80	18·55	23·81	22·56	21·31	20·06
1200	12·5	10·0	22·20	20·95	20·50	19·25	24·51	23·26	22·01	20·76
1300	11·8	10·6	22·90	21·65	21·20	19·95	25·21	23·96	22·71	21·46
1400	11·1	11·2	23·60	22·35	21·90	20·65	25·91	24·66	23·41	22·16
1500	10·5	11·9	24·40	23·15	22·70	21·45	26·71	25·46	24·21	22·96
1600	10·0	12·5	25·10	23·85	23·40	22·15	27·41	26·16	24·91	23·66
1700	9·52	13·1	25·80	24·55	24·10	22·85	28·11	26·86	25·61	24·36
1800	9·09	13·7	26·50	25·25	24·80	23·55	28·81	27·56	26·31	25·06
1900	8·70	14·4	27·30	26·05	25·60	24·35	29·61	28·36	27·11	25·86
2000	8·33	15·0	28·00	26·75	26·30	25·05	30·31	29·06	27·81	26·56
2250	7·54	16·6	29·85	28·60	28·15	26·90	32·16	30·91	29·66	28·41
2500	6·90	18·1	31·60	30·35	29·90	28·65	33·91	32·66	31·41	30·16
½ mile	6·58	19·0	32·64	31·39	30·94	29·69	34·95	33·70	32·45	31·20
3000	5·88	21·2	35·20	33·95	33·50	32·25	37·51	36·26	35·01	33·76
3250	5·48	22·8	37·05	35·80	35·35	34·10	39·36	38·11	36·86	35·61
3500	5·13	24·3	38·80	37·55	37·10	35·85	41·11	39·86	38·61	37·36
3750	4·82	25·9	40·65	39·40	38·95	37·70	42·96	41·71	40·46	39·21
4000	4·54	27·5	42·50	41·25	40·80	39·55	44·81	43·56	42·31	41·06
4250	4·30	29·1	44·35	43·10	42·65	41·40	46·66	45·41	44·16	42·91
4500	4·08	30·6	46·10	44·85	44·40	43·15	48·41	47·16	45·91	44·66
4750	3·88	32·2	47·95	46·70	46·25	45·00	50·26	49·01	47·76	46·51
5000	3·70	33·8	49·80	48·55	48·10	46·85	52·11	50·86	49·61	48·36
1 mile	3·52	35·5	51·78	50·53	50·08	48·83	54·09	52·84	51·59	50·34
1¼ m.	2·86	43·8	61·40	60·15	59·70	58·45	63·71	62·46	61·21	59·96
1½ m.	2·40	52·1	71·02	69·77	69·32	68·07	73·33	72·08	70·83	69·58
1¾ m.	2·07	60·4	80·64	79·39	78·94	77·69	82·95	81·70	80·45	79·20
2 m.	1·82	68·7	90·26	89·01	88·56	87·31	92·57	91·32	90·07	88·82

BY CARTS.—LABOR $1 PER DAY, OF 10 WORKING HOURS.

Length of lead, or distance to which the earth is hauled, in feet.	Number of cubic yards in place, hauled per day by each cart.	Cost per cubic yard in place, for hauling and emptying only.	PURE STIFF CLAY, OR CEMENTED GRAVEL. TOTAL COST PER CUBIC YARD, EXCLUSIVE OF PROFIT TO CONTRACTOR.				LIGHT SANDY SOILS. TOTAL COST PER CUBIC YARD, EXCLUSIVE OF PROFIT TO CONTRACTOR. (CENTS.)			
			Picked and Spread.	Picked and Wasted.	Ploughed and Spread.	Ploughed and Wasted.	Picked and Spread.	Picked and Wasted.	Ploughed and Spread.	Ploughed and Wasted.
Feet.	cub. yds.	cts.	cts.	cts.	cts.	cts.	cts.	cts.	cts.	cts.
25	47·0	2·66	19·00	17·75	14·50	13·25	11·52	10·77	10·25	9·50
50	44·4	2·81	19·17	17·92	14·67	13·42	11·69	10·94	10·42	9·67
75	42·1	2·97	19·36	18·11	14·86	13·61	11·88	11·13	10·61	9·86
100	40·0	3·12	19·53	18·28	15·03	13·78	12·05	11·30	10·78	10·03
150	36·4	3·43	19·89	18·64	15·39	14·14	12·41	11·66	11·14	10·89
200	33·3	3·75	20·26	19·01	15·76	14·51	12·78	12·03	11·51	10·76
300	28·6	4·37	20·98	19·73	16·48	15·23	13·50	12·75	12·23	11·48
400	25·0	5·00	21·71	20·46	17·21	15·96	14·23	13·48	12·46	12·21
500	22·2	5·63	22·44	21·19	17·94	16·69	14·96	14·21	13·69	12·94
600	20·0	6·25	23·16	21·91	18·66	17·41	15·68	14·93	14·41	13·66
700	18·2	6·87	23·88	22·63	19·38	18·13	16·40	15·65	15·13	14·38
800	16·7	7·48	24·59	23·34	20·09	18·84	17·11	16·36	15·84	15·09
900	15·4	8·12	25·23	23·98	20·73	19·48	17·75	17·00	16·48	15·73
1000	14·3	8·74	26·05	24·80	21·55	20·30	18·57	17·82	17·30	16·55
1100	13·8	9·40	26·81	25·56	22·31	21·06	19·33	18·58	18·06	17·31
1200	12·5	10·0	27·51	26·26	23·01	21·76	20·03	19·28	18·76	18·01
1300	11·8	10·6	28·21	26·96	23·71	22·46	20·73	19·98	19·46	18·71
1400	11·1	11·2	28·91	27·66	24·41	23·16	21·43	20·68	20·16	19·41
1500	10·5	11·9	29·71	28·46	25·21	23·96	22·23	21·48	20·96	20·21
1600	10·0	12·5	30·41	29·16	25·91	24·66	22·93	22·18	21·66	20·91
1700	9·52	13·1	31·11	29·86	26·61	25·36	23·63	22·88	22·36	21·61
1800	9·09	13·7	31·81	30·56	27·31	26·06	24·33	23·58	23·06	22·31
1900	8·70	14·4	32·61	31·36	28·11	26·86	25·13	24·38	23·86	23·11
2000	8·33	15·0	33·31	32·06	28·81	27·56	25·83	25·08	24·56	23·81
2250	7·54	16·6	85·16	33·91	30·66	29·41	27·68	26·93	26·41	25·66
2500	6·90	18·1	36·91	35·66	32·41	31·16	29·43	28·68	28·16	27·41
½ mile	6·58	19·0	37·95	36·70	33·45	32·20	30·47	29·72	29·20	28·45
3000	5·88	21·2	40·51	39·26	36·01	34·76	33·03	32·28	31·76	31·01
3250	5·48	22·8	42·36	41·11	37·86	36·61	34·88	34·13	33·61	32·86
3500	5·18	24·3	44·11	42·86	39·61	38·36	36·63	35·88	35·36	34·61
3750	4·82	25·9	45·96	44·71	41·46	40·21	38·48	37·73	37·21	36·46
4000	4·54	27·5	47·81	46·56	43·31	42·06	40·33	39·58	39·06	38·31
4250	4·30	29·1	49·66	48·41	45·16	43·91	42·18	41·45	40·93	40·18
4500	4·08	30·6	51·41	50·16	46·91	45·66	43·93	43·18	42·66	41·91
4750	3·88	32·2	53·26	52·01	48·76	47·51	45·78	45·03	44·51	43·76
5000	3·70	33·8	55·11	53·86	50·61	49·36	47·63	46·88	46·36	45·61
1 mile	3·52	35·5	57·09	55·84	52·59	51·34	49·61	48·86	48·34	47·59
1¼ m.	2·86	43·8	66·91	65·46	62·21	60·96	59·23	58·48	57·96	57·21
1½ m.	2·40	52·1	76·33	75·08	71·83	70·58	68·85	68·10	67·58	66·83
1¾ m.	2·07	60·4	85·95	84·70	81·45	80·20	78·47	77·72	77·20	76·45
2 m.	1·82	68·7	95·57	94·82	91·07	89·82	88·09	87·34	86·82	86·07

ARTICLE 10.

BY WHEELBARROWS.—The cost by barrows may be estimated in the same manner as by carts. See Articles 1, &c. Men in wheeling move at about the same average rate as horses do in hauling, that is, 2½ miles an hour, or 200 feet per minute, or 1 minute per every 100-feet length of lead. The time occupied in loading, emptying, &c. (when, as is usual, the wheeler loads his own barrow,) is about 1·25 minutes, without regard to length of lead; beside which, the time lost in occasional short rests, in adjusting the wheeling-plank, and in other incidental causes, amounts to about $\frac{1}{10}$ part of his whole time; so that we must in practice consider him as actually working but 9 hours out of his 10 working ones, at the rate of 2·25 minutes per 100 feet of lead. To find, then, the number of barrow-loads which he can remove in a day, multiply the number of minutes (600) in a working day by ·9; and divide the product by the sum of 1·25, added to the number of 100-feet lengths in the lead; that is,

$$\frac{\text{The number of minutes in a working day} \times \cdot 9}{1\cdot 25 + \text{the number of 100-feet lengths of lead}} = \text{the number of trips or of loads removed per day per barrow.}$$

Remark. For *rock*, which requires more time for loading, say

$$\frac{\text{No. of minutes in a working day} \times \cdot 9}{1\cdot 6 + \text{No. of 100-feet lengths of lead}} = \frac{\text{No. of loads removed}}{\text{per day, per barrow}}$$

The number of loads divided by 14 will give the number of cubic yards, since a cubic yard, measured in place, averages about 14 loads. And the cost of a wheeler and barrow per day, (say $1 per man, and 5 cents per barrow,) divided by the number of cubic yards, will give the cost per yard for loading, wheeling, and emptying.

Example. How many cubic yards of common loam, measured in place, will one man load, wheel, and empty, per day of 10 working hours (or 600 minutes); the lead, or distance to which the earth is removed being 1000 feet (or 10 lengths of 100 feet); and what will be the expense per yard, supposing the laborer and barrow to cost $1.05 per day?

Here, $\dfrac{600 \text{ minutes} \times \cdot 9}{1\cdot 25 + 10 \text{ lengths}} = \dfrac{540}{11\cdot 25} = 48$ trips, or loads per day.

And $\dfrac{48}{14} = 3\cdot 43$ cubic yards per day. And $\dfrac{105 \text{ cents}}{3\cdot 43 \text{ cub. yds.}} = 30\cdot 6$ cts.

per cubic yard for loading, wheeling away, emptying, and returning. This would be increased almost inappreciably by the cost of the shovel, which, in the following tables, however, is included in the cost of tools.

ARTICLE 11.

The following tables are calculated as in the case of carts, by first finding columns 2 and 3 by means of the Rule in Article 10, and then adding to each sum in column 3, the variable quantity of ·1 of a cent per cubic yard per 100 feet of lead for keeping the wheeling-planks in order; and the prices of loosening, spreading, superintendence, water-carrying, &c., per cubic yard, as given in the preceding Articles 2 to 7.

BY WHEELBARROWS.—LABOR $1 PER DAY, OF 10 WORKING HOURS.

Length of load, or distance to which the earth is wheeled, in feet.	Number of cubic yards in place, loaded, and wheeled per day; each barrow.	Cost per cubic yard in place, for loading, wheeling, and emptying.	COMMON LOAM.				STRONG, HEAVY SOILS.			
			TOTAL COST PER CUBIC YARD, EXCLUSIVE OF PROFIT TO CONTRACTOR. (CENTS.)				TOTAL COST PER CUBIC YARD, EXCLUSIVE OF PROFIT TO CONTRACTOR. (CENTS.)			
			Picked and Spread.	Picked and Wasted.	Ploughed and Spread.	Ploughed and Wasted.	Picked and Spread.	Picked and Wasted.	Ploughed and Spread.	Ploughed and Wasted.
Feet.	cub. yds.	cts.	cts.	cts.	cts.	cts.	cts.	cts.	cts.	cts.
25	25·7	4·09	10·12	8·87	8·42	7·17	11·62	10·37	9·12	7·87
50	22·1	4·75	10·80	9·55	9·10	7·85	12·30	11·05	9·80	8·55
75	19·8	5·44	11·52	10·27	9·82	8·57	13·02	11·77	10·52	9·27
100	17·1	6·14	12·24	10·99	10·54	9·29	13·74	12·49	11·24	9·99
150	14·0	7·50	13·65	12·40	11·95	10·70	15·15	13·90	12·65	11·40
200	11·9	8·82	15·02	13·77	13·32	12·07	16·52	15·27	14·02	12·77
250	10·3	10·2	16·45	15·20	14·75	13·50	17·95	16·70	15·45	14·20
300	9·07	11·6	17·90	16·65	16·20	14·95	19·40	18·15	16·90	15·65
350	8·14	12·9	19·25	18·00	17·55	16·30	20·75	19·50	18·25	17·00
400	7·36	14·3	20·70	19·45	19·00	17·75	22·20	20·95	19·70	18·45
450	6·71	15·6	22·05	20·80	20·35	19·10	23·55	22·30	21·05	19·80
500	6·17	17·0	23·50	22·25	21·80	20·55	25·00	23·75	22·50	21·25
600	5·82	19·7	26·30	25·05	24·60	23·35	27·80	26·55	25·30	24·05
700	4·67	22·5	29·20	27·95	27·50	26·25	30·70	29·45	28·20	26·95
800	4·17	25·2	32·00	30·75	30·30	29·05	33·50	32·25	31·00	29·75
900	3·76	27·9	34·80	33·55	33·10	31·85	36·30	35·05	33·80	32·55
1000	3·43	30·6	37·60	36·35	35·90	34·65	39·10	37·85	36·60	35·35
1200	2·91	36·1	43·30	42·05	41·60	40·35	44·80	43·55	42·30	41·05
1400	2·53	41·5	48·90	47·65	47·20	45·95	50·40	49·15	47·90	46·65
1600	2·24	46·9	54·50	53·45	52·80	51·55	56·00	54·75	53·50	52·25
1800	2·00	52·5	60·30	59·05	58·60	57·35	61·80	60·55	59·30	58·05
2000	1·81	58·0	66·00	64·75	64·30	63·05	67·50	66·25	65·00	63·75
2200	1·66	63·3	71·50	70·25	69·80	68·55	73·00	71·75	70·50	69·25
2400	1·53	68·6	77·00	75·75	75·30	74·05	78·50	77·25	76·00	74·75
½ mile	1·39	75·5	84·14	82·89	82·44	81·19	85·64	84·39	83·14	81·89

BY WHEELBARROWS.—LABOR $1 PER DAY, OF 10 WORKING HOURS.

Length of lead, or distance to which the earth is wheeled.	Number of cubic yards in place, loaded, and wheeled per day; each barrow.	Cost per cubic yard in place, for loading, wheeling, and emptying.	PURE STIFF CLAY, OR CEMENTED GRAVEL. TOTAL COST PER CUBIC YARD, EXCLUSIVE OF PROFIT TO CONTRACTOR.				LIGHT SANDY SOILS. TOTAL COST PER CUBIC YARD, EXCLUSIVE OF PROFIT TO CONTRACTOR.			
			Picked and Spread.	Picked and Wasted.	Ploughed and Spread.	Ploughed and Wasted.	Picked and Spread.	Picked and Wasted.	Ploughed and Spread.	Ploughed and Wasted.
Feet.	cub. yds.	cts.	cts.	cts.	cts.	cts.	cts.	cts.	cts.	cts.
25	25·7	4·09	14·62	13·37	10·12	8·87	8·79	8·04	7.52	6·77
50	22·1	4·75	15·30	14·05	10·80	9·55	9·47	8·72	8·20	7·45
75	19·3	5·44	16·02	14·77	11·52	10·27	10·19	9·44	8·92	8·17
100	17·1	6·14	16·74	15·49	12·24	10·99	10·91	10·16	9·64	8·89
150	14·0	7·50	18·15	16·90	13·65	12·40	12·32	11·57	11·05	10·30
200	11·9	8·82	19·52	18·27	15·02	13·77	13·69	12·94	12·42	11·67
250	10·3	10·2	20·95	19·70	16·45	15·20	15·12	14·37	13·85	13·10
300	9·07	11·6	22·40	21·15	17·90	16·65	16·57	15·82	15·30	14·55
350	8·14	12·9	23·75	22·50	19·25	18·00	17·92	17·17	16·65	15·90
400	7·36	14·3	25·20	23·95	20·70	19·45	19·87	18·62	18·10	17·35
450	6·71	15·6	26·55	25·30	22·05	20·80	20·72	19·97	19·45	18·70
500	6·17	17·0	28·00	26·75	23·50	22·25	22·17	21·42	20·90	20·15
600	5·32	19·7	30·80	29·55	26·30	25·05	24·97	24·22	23·70	22·95
700	4·67	22·5	33·70	32·45	29·20	27·95	27·87	27·12	26·60	25·85
800	4·17	25·2	36·50	35·25	32·00	30·75	30·67	29·92	29·40	28·65
900	3·76	27·9	39·30	38·05	34·80	33·55	33·47	32·72	32·20	31·45
1000	3·43	30·6	42·10	40·85	37·60	36·35	36·27	35·52	35·00	34·25
1200	2·91	36·1	47·80	46·55	43·30	42·05	41·97	41·22	40·70	39·90
1400	2·53	41·5	53·40	52·15	48·90	47·65	47·57	46·82	46·30	45·55
1600	2·24	46·9	59·00	57·75	54·50	53·25	53·17	52·42	51·90	51·15
1800	2·00	52·5	64·80	63·55	60·30	59·05	58·97	58·22	57·70	56·95
2000	1·81	58·0	70·50	69·25	66·00	64·75	64·67	63·92	68·40	62·65
2200	1·66	63·3	76·00	74·75	71·50	70·25	70·17	69·42	68·90	68·15
2400	1·53	68·6	81·50	80·25	77·00	75·75	75·67	74·92	74·40	73·65
½ mile	1·39	75·5	88·64	87·39	84·14	82·89	82·81	82·06	81·54	80·79

ARTICLE 12.

BY WHEELED SCRAPERS AND DRAG SCRAPERS.—The body of the wheeled scraper is a box of smooth sheet-steel about $3\frac{1}{2}$ feet square by 15 inches deep, containing about $\frac{1}{2}$ cubic yard of earth when "even full." The box is open in front (in some machines it is closed by an "end gate" when full), and can be raised and lowered, and revolved on a horizontal axis. To fill the box, it is lowered into, and held down in, the earth, while the team draws the machine forward. When full, it is raised to about a foot above ground; and, on reaching the dump, is unloaded by being overturned on its axis. All the movements of the box are made by means of levers, and without stopping the team, which thus travels constantly. The wheels have broad tires, to prevent them from cutting into the ground.

In the *drag* scraper the box, owing to the greater resistance to traction, is made much smaller. It contains about .15 to .25 cubic yard in place, and is always open in front. The operation of the drag scraper is similar to that of the wheeled scraper, except that the box, when filled, rests upon the ground and is dragged over it by the team.

Each scraper ("wheeled" or "drag") requires the constant use of a team of two horses with a driver. Besides, a number of men, depending on the shortness of the lead and the number of scrapers, are required in the pit and at the dump, to load the scrapers (by holding the box down into the earth) and unload them (by tipping the box). Except in sand, or in very soft soil, it is economical to use a plow before scraping.

The severest work for the team is the filling of the box; and this occurs oftenest where the lead is shortest. Hence smaller scrapers are used on short than on long hauls. We base our calculations on the following loads:

For drag scrapers (used only on short hauls)............ .2 cubic yard
For wheeled scrapers
    lead less than 100 feet.. .33 "
    " 100 to 300 feet... .4 "
    " 400 to 500 feet... .5 "
    " over 500 feet... .6 "

The daily expense per scraper, for driver's wages and the use of a 2-horse team, is about $3.50. For leads of 400 feet and over, we add 50 cents per day for use of "snatch team" to help load the larger scrapers then used. One snatch team generally serves a number of scrapers.

Owing to the fact that the teams are constantly in motion without rest, they travel somewhat more slowly than with carts. We take 150 feet per minute (or 75 feet of *lead* per minute) as an average.

In loading and unloading, the teams not only go out of their way in order to turn around, but travel more slowly than when simply hauling. To cover this we make an addition of 25 feet to each length of lead, whether long or short, for wheeled scrapers; and 15 feet for drag scrapers.

We add 1 cent per cubic yard for the cost of loading and dumping the scrapers; and estimate the approximate cost of the other items as follows:

Repairs of cart road $\frac{1}{10}$ cent per cubic yard in place for each 100 feet of lead.

Loosening.	Light Soils. Cts. per cub. yd. in place.	Heavy Soils. Cts. per cub. yd. in place.
by pick	*	5.
by shovel	*	2.
Spreading	1.	1.5
Superintendence, wear and tear, etc	1.	1.

We repeat that our figures are to be regarded merely as tolerable approximations, and subject to great variations according to skill of contractor and superintendent, strength of teams, character of material moved, state of weather, etc., etc.

$$\frac{\text{No. of trips per day}}{\text{per } \textit{wheeled} \text{ scraper}} = \frac{\text{No. (600) of minutes in a working day}}{\text{No. of 75 ft. lengths in (lead + 25 ft.)}}$$

$$\frac{\text{No. of trips per day}}{\text{per } \textit{drag} \text{ scraper}} = \frac{\text{No. (600) of minutes in a working day}}{\text{No. of 75 ft. lengths in (lead + 15 ft.)}}$$

$$\frac{\text{No. of cub. yds. in place moved}}{\text{per day by each scraper}} = \frac{\text{No. of trips per}}{\text{day per scraper}} \times \frac{\text{No. of cub. yds. in place,}}{\text{per scraper per trip.}}$$

$$\frac{\text{Cost per cub. yd. in place,}}{\text{for loading, hauling,}}_{\text{dumping and returning}} = \frac{\text{Daily expense of one scraper}}{\text{No. of cub. yds. in place, moved}}_{\text{per day by each scraper}} + \frac{1 \text{ cent for loading}}{\text{and dumping}}$$

$$\frac{\text{Total cost per}}{\text{cubic yard in}}_{\text{place exclusive}} = \frac{\text{Cost per cub. yd. in}}{\text{place for loading,}}_{\text{hauling, dumping,}} + \frac{.1 \text{ cent per cub. yd.}}{\text{in place for each}}_{\text{100 feet of lead, for}} + \frac{\text{Cost, per cub. yd. in place,}}{\text{of loosening, spreading or}}_{\text{wasting, and superintend-}}$$
$$\text{of contractor's}_{\text{profit}} \quad \text{and returning} \quad \text{repairs of road.} \quad \text{ence &c.}$$

BY WHEELED SCRAPERS.
Labor $1 per day of 10 working hours.

Length of lead, or dist. to which earth is hauled.	Quantity in place, hauled per day by each scraper.	Cost per cub. yd. in place, for loading, hauling, dumping, and returning.	Total cost, per cub. yd. in place, exclusive of contractor's profit.					
			Light Soils.		Heavy Soils.			
					Picked and		Plowed and	
			Spread.	Wasted.	Spread.	Wasted.	Spread.	Wasted.
Feet.	cub. yds.	cts.	cts.	cts.	cts.	cts.	cts.	cts.
50	200	2.8	4.9	3.9	10	8.5	7.3	5.8
100	140	3.4	5.5	4.5	11	9.5	8	6.5
150	105	4.3	6.5	5.5	12	11	9	7.5
200	80	5.4	7.6	6.6	13	12	10	8.5
300	56	7.3	9.6	8.6	15	14	12	11
400	50	8.5	11	10	16	15	13	12
600	43	10	13	12	18	17	15	14
800	33	13	16	15	21	20	18	17
1000	27	16	19	18	25	24	22	21

*Light soils can generally be advantageously loosened by the scrapers themselves in the act of loading.

BY DRAG SCRAPERS.
Labor $1 per day of 10 working hours.

Length of lead, or distance to which earth is hauled.	Quantity in place, hauled per day by each scraper.	Cost per cub. yd in place, for loading, hauling, dumping and returning.	Total cost, per cubic yard in place, exclusive of contractor's profit.					
			Light Soils.		Heavy Soils.			
					Picked and		Plowed and	
			Spread.	Wasted.	Spread.	Wasted.	Spread.	Wasted.
Feet.	Cu. yds.	Cts.	Cts.	Cts.	Cts.	Cts.	Cts.	Cts.
less than 40	220	2.6	4.6	3.6	10	8.5	7	5.5
50	140	3.5	5.5	4.5	11	9.5	8	6.5
75	100	4.5	6.6	5.6	12	11	9	8
100	80	5.4	7.5	6.5	13	12	10	9
150	54	7.5	9.6	8.6	15	14	12	11
200	42	9.3	12.	11.	17	16	14	13

Both wheeled and drag scrapers are made by Western Wheel Scraper Co., Mount Pleasant, Iowa; by Kilbourne & Jacobs Manufacturing Co., Columbus, Ohio; by Fay Manufacturing Co., Elyria, Ohio, and others. A medium-sized wheeled scraper, weighing 450 ℔s, and carrying ·4 cubic yard, costs about from $50 to $70. A drag scraper weighs about 100 ℔s., and cost about $14.

ARTICLE 13.

BY CARS AND LOCOMOTIVE, ON LEVEL TRACK, based upon the following assumptions. Trains of 10 cars, each car containing 1½ cubic yards of earth measured in place. Average speed of trains, including starting and stopping, but not standing, 10 miles per hour, = 5 miles of *lead* per hour. Labor $1 per day of 10 working hours. Loosening, loading (by shovelers), spreading, wear etc. of tools, superintendence, etc., the same as with carts, Arts. 2, 3, 5, and 7. Loss of time in each trip for loading, unloading, etc., 9 minutes, = ·15 hour. Therefore,

$$\text{Number of trips per day per train} = \frac{\text{The number (10) of hours in a working day}}{\cdot 15 + \text{the number of 5-mile lengths in the lead.}}$$

$$\text{Number of cubic yards in place, per day per train.} = \text{Number of trips per day} \times \text{Number (10) of cars in a train.} \times \text{Number (1·5) of cubic yards in place in each car.}$$

$$\text{Cost per cubic yard, in place, for hauling, dumping, and returning} = \frac{\text{One day's train expenses} + \text{one day's cost of track}}{\text{Number of cubic yards in place per day per train.}}$$

One day's train expenses:

```
Cost of 10 cars @ $100..........................................$1,000
   "   locomotive.................................................. 3,000
                                                              ———— $4,000
One day's interest, at 6 per cent. on cost of train,................ $0.67
Wages of engine driver (who fires his own engine).................. 2.00
   "    foreman at dump ........................................... 2.00
   "    3 men at dump at $1 ....................................... 3.00
Fuel............................................................... 2.00
Water.............................................................. 1.00
Repairs of locomotive and cars..................................... 2.33
                                                                    ——————
            Total daily expense of one train........................$13.00
```

The daily expense of track, for interest and repairs, may be taken at $3 for each mile, or fraction of a mile, of lead. Therefore,

$$\left.\begin{array}{l}\text{Cost per cubic yard in place,}\\ \text{for hauling, dumping, and}\\ \text{returning}\end{array}\right\} = \frac{\$13 + (\$3 \text{ for each mile of lead})}{\underset{\text{trips per day}}{\text{Number of}} \times \underset{\text{of cars in a}}{\text{Number (10)}} \times \underset{\text{of cubic yards}}{\text{Number (1·5)}}}{\text{per train} \quad\quad \text{train} \quad\quad \text{in each car}}$$

$$\left.\begin{array}{l}\text{Total cost per cubic}\\ \text{yard in place, exclu-}\\ \text{sive of contractor's}\\ \text{profit}\end{array}\right\} = \begin{array}{l}\text{Cost per cubic yard}\\ \text{in place for hauling,}\\ \text{dumping, and return-}\\ \text{ing}\end{array} + \begin{array}{l}\text{Cost per cubic yard, in place, for}\\ \text{loosening, loading, spreading or}\\ \text{wasting, and superintendence,}\\ \text{&c. (Arts. 2, 3, 5, and 7.)}\end{array}$$

BY CARS AND LOCOMOTIVE.

Labor $1 per day of 10 working hours.

Length of lead, or distance to which the earth is hauled.	Number of cubic yards, in place, hauled per day by each train.	Cost per cubic yard, in place, for hauling, dumping, and returning.	Total cost per cubic yard, in place, exclusive of contractor's profit.							
			Light Sandy Soils.				Strong Heavy Soils.			
			Picked and Spread.	Picked and Wasted.	Ploughed and Spread.	Ploughed and Wasted.	Picked and Spread.	Picked and Wasted.	Ploughed and Spread.	Ploughed and Wasted.
Miles.	Cu. yds.	Cts.	Cts.	Cts.	Cts.	Cts.	Cts.	Cts.	Cts.	Cts.
1	4350	.4	9.7	8.4	8.4	7.2	13.7	12.4	11.3	10.
2	2700	.7	10.	8.8	8.8	7.5	14.	12.8	11.6	10.4
3	1950	1.1	10.4	9.2	9.2	7.9	14.5	13.3	12.1	10.9
4	1500	1.7	11.	9.7	9.7	8.5	15.	13.7	12.6	11.3
5	1200	2.3	11.6	10.4	10.4	9.1	15.6	14.4	13.2	12.
6	1050	3.	12.3	11.	11.	9.8	16.3	15.	13.9	12.6
7	900	3.8	13.1	11.8	11.8	10.6	17.1	15.8	14.7	13.4
8	750	4.9	14.2	13.	13.	11.7	18.2	17.	15.8	14.6
10	600	7.2	16.5	15.2	15.2	14.	20.5	19.2	18.	16.8

STEAM EXCAVATORS.

Where large amounts of work are to be done, the steam excavator, land dredge or steam shovel generally economizes time and money. Where the depth of cutting is less than 10 feet, so much time is lost in moving from place to place that the excavators do not work to advantage. In stiff soils, cuttings may be made about from 17 to 20 feet deep without changing the level of the machine. For greater depths in such soils the work is done in two levels, since the bucket or dipper cannot reach so high. But in sand and loose gravel, much deeper cuts may be made from a single level.

The excavator resembles a dredging machine in its appearance and operation. A large plate-steel bucket, like a dredging bucket, with a flat hinged bottom, and provided with steel cutting teeth, is forced into and dragged through the earth by steam power. It dumps its load, by means of the hinged bottom, either into cars for transportation, or upon the waste bank, as desired.

Each machine is mounted on a car of standard gauge, which can be coupled in an ordinary freight train. The car is made of wood or iron, as desired, and is provided with a locomotive attachment, by which it can be moved from point to point as the work proceeds. The machines can be used as wrecking or derrick cars. Each machine has a water tank, holding from 300 to 550 gallons, for the supply of its boiler.

Before beginning to excavate, the end of the car nearest the work is lifted from the track by hydraulic or screw jacks, upon which it rests while working.

In stiff soils the excavator leaves the sides of the cut nearly vertical; and the desired slope is afterwards given by pick and shovel. When the soil is hard or much frozen, it may be loosened by blasting in advance of the excavator.

Steam excavators are made by Osgood Dredge Co., Albany, N. Y.; by John Souther & Co., (the "Otis" excavator) Boston, Mass.; by Vulcan Iron Works, Toledo, O.; by Industrial Works, Bay City, Mich.; and by Pound Manufacturing Co., Lockport, N. Y.

The Osgood is made in two sizes. In No. 1 the car is 34 ft. x 10 ft., and its floor is 4 ft. above the rails. It has a four-wheeled truck near each end. The dipper holds 2 cubic yards, struck measure. The machine weighs, complete, about 40 tons, and costs about $7500 on track at works (Albany, N. Y.). In the No. 2 machine, the car is 28 ft. x 10 ft., floor 5 ft. above the rails. It has two pairs of wheels, 16 feet apart from centre to centre of axles. The dipper holds $1\frac{1}{2}$ cubic yards, struck measure. The machine weighs, complete, about 28 tons, and costs about $6000.

The excavator has to be moved forward (as the work advances) about 8 feet at a time. As regularly made, it can dig at a distance of 17 feet, horizontally, from the center of the car in any direction, and can dump 12 feet above the track. In sand or gravel it takes out, while actually digging, 3 dipperfuls (=4½ to 6 cubic yards in the dipper,=3.75 to 5 cubic yards in place) per minute; in stiff clay, 2 dipperfuls per minute (=3 to 4 cubic yards in the dipper,=2.5 to 3.33 cubic yards in place). An average day's work (10 hours) for a "No. 1" machine, including time lost in moving the machine, &c., is about 500 cubic yards in "hard-pan," and from 1200 to 1500 in sand and gravel. This allows for the usual and generally unavoidable delays in having cars ready for the excavator.

The excavators carry about 80 to 90 lbs. of steam. They burn from 100 to 150 lbs. of good hard or soft coal per hour; and require one engineer, one fireman, one cranesman, and 5 to 10 pitmen, including a boss. The pitmen are laborers, who attend to the jacks, lay track for the excavator and for the dump cars, assist in moving the latter, bring or pump water, &c., &c.

After reaching the site of the work, about 30 minutes are required for getting the excavator into working condition; and an equal length of time, after completion of the work, in getting it ready for transportation.

The following figures are taken from the records of work done by a No. 1 machine, from May to November, 1883. The material was hard clay with pockets of sand. The expenses per day of 12 working hours, at $1.50 per such day for labor, were

```
Water, (a very high allowance)..............................$ 5.00
Coal, 1½ tons bituminous.................................... 10.00
Wages of engineer.............................................. 4.00
  "     "  fireman................................................ 1.50
  "     "  cranesman or dipper-tender................... 2.50
  "     "  pit boss.............................................. 3.00
  "     "  8 pitmen at $1.50................................ 12.00
Oil, waste, repairs &c. (estimated)....................... 5.00
Interest on cost ($7500) of machine..................... 1.25
                                                           ——— $44.25
```

Reduced to our standard of $1 for labor per day of 10 working hours, this would be say $30.00 per day. Reduced to the same standard, and allowing for the greater proportional loss of time in stopping at evening and starting in the morning: the average daily quantity excavated, measured in place, was, in shallow cutting, 530 cubic yards; in deep cutting, 1200 cubic yards; average of whole operation, 800 cubic yards. This would make the cost, per cubic yard measured in place, for loosening and loading into cars, 5.67 cts., 2.5 cts., and 3.75 cts. respectively; while the cost by ploughing and shoveling, in strong heavy soils, by Arts. 2 and 3, is 7.4 cts.; and by picking and shoveling, say 10 cts.

ARTICLE 14.

REMOVING ROCK EXCAVATION BY WHEELBARROWS.—A cubic yard of hard rock, *in place*, or before being blasted, will weigh about 1·8 tons, if sandstone or conglomerate (150 lbs. per cubic foot); or 2 tons if good compact granite, gneiss, limestone, or marble (168 lbs. per cubic foot). So that, near enough for practice in the case before us, we may assume the weight of any of them to be about 1·9 tons, or 4256 lbs. per cubic yard, in place; or 158 lbs. per cubic foot.

Now, a solid cubic yard of any of these, when broken up by blasting for removal by wheelbarrows or carts, will occupy a space of about 1·8, or 1¾ cubic yards; whereas average earth, when loosened, swells to but about 1·2, or 1⅕ of its original bulk in place; although, after being made into embankment, it eventually shrinks into less than its original bulk. In estimating for earth, it is assumed that $\frac{1}{14}$ cubic yard, in place, is a fair load for a wheelbarrow. Such a cubic yard will weigh on an average 2430 lbs., or 1·09 tons; therefore, $\frac{2430}{14} =$ 174 lbs., is the weight of a barrow-load, of 2·31 cubic feet of loose earth. Assuming that a barrow of loose rock should weigh about the same as one of earth, we may take it at $\frac{1}{24}$ of a cubic yard; which gives $\frac{4256}{24} = 177$ lbs. per load of loose rock, occupying 2 cubic feet of space.

In the following table, columns 2 and 3 are prepared on the same principle as for earth, as directed in Article 10. Column 4 is made up by adding to each amount in column 3, ·2 of a cent for each 100 feet length of lead, for keeping the wheeling-planks in order; and 45 cents per cubic yard, in place, as the actual cost for loosening, including tools, drilling, powder, &c.; as well as moderate drainage, and every ordinary contingency not embraced in column 3. Contractor's profits, of course, are not here included.

Ample experience shows that when labor is at $1 per day, the foregoing 45 cents per cubic yard, in place, is a sufficiently liberal allowance for loosening hard rock under all ordinary circumstances. In practice it will generally range between 30 and 60 cents; depending on the position of the strata, hardness, toughness, water, and other considerations. Soft shales, and other allied rocks, may frequently be loosened by pick and plough, as low as 15 to 20 cents; while, on the other hand, shallow cuttings of very tough rock, with an unfavorable position of strata, especially in the bottoms of excavations, may cost $1, or even considerably more. These, however, are exceptional cases, of comparatively rare occurrence. The quarrying of average hard rock requires about ¼ to ⅓ lb. of powder per cubic yard, in place; but the nature of the rock, the position of the strata, &c., may increase it to ½ lb., or more. Soft rock frequently requires more powder than hard. A good churn-driller will drill from 8 to 12 feet in depth, of holes about 2½ feet deep, and 2 inches diameter, per day, in average hard rock, at from 12 to 18 cents per foot. Drillers receive higher wages than common laborers.

HARD ROCK, BY WHEELBARROWS.

Labor $1 per day of 10 working hours.

Length of lead, or distance to which the rock is wheeled.	Number of cubic yards, in place, wheeled per day by each barrow.	Cost per cubic yard, in place, for loading, wheeling, and emptying.	Total cost per cubic yard, in place, exclusive of profit to contractor.	Length of lead, or distance to which the rock is wheeled.	Number of cubic yards, in place, wheeled per day by each barrow.	Cost per cubic yard, in place, for loading, wheeling, and emptying.	Total cost per cubic yard, in place, exclusive of profit to contractor.
Feet.	cubic yds.	cents.	cents.	Feet.	cubic yds.	cents.	cents.
25	12·2	8·64	53·7	600	2·96	35·5	81·7
50	10·7	9·81	54·9	700	2·62	40·1	86·5
75	9·58	11·0	56·2	800	2·34	44·8	91·4
100	8·66	12·1	57·3	900	2·12	49·5	96·3
150	7·26	14·5	59·8	1000	1·94	54·1	101·1
200	6·25	16·8	62·2	1200	1·65	63·6	115·0
250	5·49	19·1	64·6	1400	1·44	72·9	120·7
300	4·89	21·5	67·1	1600	1·28	82·2	130·4
350	4·41	23·8	69·5	1800	1·15	91·5	140·1
400	4·02	26·1	71·9	2000	1·04	100·8	149·8
450	3·69	28·5	74·4	2200	·953	110·2	159·6
500	3·41	30·8	76·8	2400	·879	119·5	169·3

ARTICLE 15.

REMOVING ROCK EXCAVATION BY CARTS.—A cart-load of rock may be taken at ⅙ of a cubic yard, in place. This will weigh, on an average, 851 lbs.; or but 41 lbs. more than a cart-load of average soil. Since the cart itself will weigh about ½ a ton, the total loads are very nearly equal in both cases. Columns 2 and 3 of the following table are prepared on the same principle as for earth, as directed in Article 4. Column 4 is made up by adding to each amount in column 3, the following items. For blasting, (and for everything except those in column 3; loading, and repairs of cart-road,) 45 cents per cubic yard, in place; for loading, 8 cents, per cubic yard, in place; and for repairs of road, ·2, or ⅕ of a cent for each 100-feet length of lead. Contractor's profit not included.

HARD ROCK, BY CARTS.

Labor $1 per day, of 10 working hours.

Length of lead, or distance to which the rock is hauled.	Number of cubic yards, in place, hauled per day, by each cart.	Cost per cubic yard, in place, for hauling, and emptying.	Total cost per cubic yard, in place, exclusive of profit to contractor.	Length of lead, or distance to which the rock is hauled.	Number of cubic yards, in place, hauled per day, by each cart.	Cost per cubic yard, in place, for hauling, and emptying.	Total cost per cubic yard, in place, exclusive of profit to contractor.
Feet.	cubic yds.	cents.	cents	Feet.	cubic yds.	cents.	cents.
25	19·2	6·51	59·6	1800	5·00	25·0	81·6
50	18·5	6·77	59·9	1900	4·80	26·0	82·8
75	17·8	7·03	60·2	2000	4·62	27·1	84·1
100	17·1	7·29	60·5	2250	4·21	29·7	87·2
150	16·0	7·81	61·1	2500	3·87	32·3	90·3
200	15·0	8·33	61·7	½ mile	3·70	33·7	92·0
300	13·3	9·37	63·0	3000	3·33	37·5	96·5
400	12·0	10·4	64·2	3250	3·12	40·1	99·6
500	10·9	11·5	65·5	3500	2·92	42·8	102·8
600	10·0	12·5	66·7	3750	2·76	45·3	105·8
700	9·23	13·6	68·0	4000	2·61	47·9	108·9
800	8·57	14·6	69·2	4250	2·47	50·6	112·1
900	8·00	15·6	70·4	4500	2·35	53·2	115·2
1000	7·50	16·7	71·7	4750	2·24	55·8	118·3
1100	7·06	17·7	72·9	5000	2·14	58·4	121·4
1200	6·67	18·7	74·1	1 mile	2·04	61·2	124·8
1300	6·32	19·8	75·4	1¼ "	1·67	75·0	141·2
1400	6·00	20·8	76·6	1½ "	1·41	88·8	157·6
1500	5·71	21·9	77·9	1¾ "	1·22	102·5	174·0
1600	5·45	22·9	79·1	2 "	1·08	116·3	190·4
1700	5·22	24·0	80·4	2¼ "	·962	130·0	206·8

What is called "loose rock" will cost about 30 cts. per yard less than the prices in the last two tables; and even *solid* rock will *average* about 10 cts. per yard less than the tabular prices, at the foregoing rates of labor; the difference in both cases being in the item of loosening alone.

ARTICLE 16.

REMOVING ROCK EXCAVATION BY CARS AND LOCOMOTIVE, on level track. Calculations based upon the following assumptions. Trains of 10 cars, each car containing 1 cubic yard of rock measured in place. Average speed of trains, including starting and stopping, but not standing, 10 miles per hour = 5 miles of *lead* per hour. Labor $1 per day of 10 working hours. Loosening, 45 cts. per cubic yard in place. Loading, 8 cts. per cubic yard in place. Cost of track, for interest and repairs, $3 per day per mile of lead. The calculations are the same, in principle, as those in Art. 13.

HARD ROCK, BY CARS AND LOCOMOTIVE.
Labor $1 per day of 10 working hours.

Length of lead, or distance to which the rock is hauled.................... miles	1	3	5	7	10
Number of cubic yards, in place, hauled per day by each train...........................	2900	1300	800	600	400
Cost, per cubic yard in place, for hauling, dumping and returning.................cents	.6	1.7	3.5	5.7	10.8
Total cost, per cubic yard in place, exclusive of contractor's profit.................cents	53.6	54.7	56.5	58.7	63.8

www.ingramcontent.com/pod-product-compliance
Lightning Source LLC
Chambersburg PA
CBHW020256090426
42735CB00009B/1109